Memoirs Of John Martyn And Of Thomas Martyn: Professors Of Botany In The University Of Cambridge

George Cornelius Gorham

MEMOIRS

OF

JOHN MARTYN, F.R.S.,

AND OF

THOMAS MARTYN, B.D., F.R.S.,F.L.S.,

Professors of Botany

IN THE UNIVERSITY OF CAMBRIDGE.

———

BY

GEORGE CORNELIUS GORHAM, B.D.,

LATE FELLOW OF QUEEN'S COLLEGE, CAMBRIDGE.

———

LONDON:

HATCHARD AND SON, PICCADILLY ;

DEIGHTON, CAMBRIDGE ; EMERY, ST. NEOT'S ; AND IBBS, KIMBOLTON.

—

MDCCCXXX.

TO

THE REV. JOHN KING MARTYN, M. A.,

OF PERTENHALL, BEDFORDSHIRE,

THESE MEMOIRS

OF

TWO MEMBERS OF HIS FAMILY,

NOT LESS DISTINGUISHED

BY

THEIR PRIVATE VIRTUES

THAN BY

THEIR SCIENTIFIC PURSUITS,

ARE RESPECTFULLY INSCRIBED BY

HIS AFFECTIONATE SON-IN-LAW

G. C. GORHAM.

PREFACE.

The Compiler of the following Memoirs does not pretend to offer any apology for having added a volume to the stock of our national Biography. If he have but performed his task with moderate care, he may, without presumption, calculate on the thanks of many readers, who, he may fairly expect, will be interested in an account of the lives and writings of two eminent British Botanists. There are not many instances in which it can be said, that two individuals, members of the same family, cultivated the same science with distinguished ability, for more than a century*;—perhaps there is none in which it can be added, that the Father transmitted to the

* For 107 years—from 1718 to 1825.

Son his public honors as well as his private
talents; both having, in succession, and in
the same University, adorned the Chair appropriated to their favourite science, during the
long period of ninety-three years![b] Such was
the unusual course of events, with regard to
the two Professors, JOHN and THOMAS MAR-
TYN. By their lectures, and especially by
their writings, these indefatigable men contributed largely to the advancement of Botanical
knowledge in this country:—in fact, an account of the two MARTYNS constitutes no
inconsiderable portion of the history of the
science to which they were devoted, as cultivated, in England, during the whole of the
eighteenth century.

The former of these Memoirs was published
nearly sixty years ago, in a small piece entitled, " *Some account of the late* JOHN MAR-
TYN, F. R. S., *and his writings.* 12mo. pp. 63.

[b] From 1732 to 1825.

London, 1770." It was prefixed to a posthumous volume, mentioned below, (pp. 67, 138,)
and was drawn up by Professor Thomas
Martyn, soon after his learned father's death,
as " a tribute to his memory." It is now
republished, with considerable additions, distinguished from the original matter by being
inserted between brackets. A few paragraphs
are transposed; others, which had a merely
temporary reference, are omitted; and an
Appendix, which was nearly as long as the
original Memoir itself, has been incorporated
with the text and notes.

The Memoir of Professor THOMAS MARTYN,
is now for the first time given to the public.
The materials from which it has been compiled
are chiefly the following :—

1. A short Auto-biographical Manuscript,
written by the Professor in his eighty-sixth
year, and comprised in 53 quarto pages.
This has occasionally been cited in the very

words of the venerable writer, (see p. **262**) but it has been thought advisable, generally, to make a free use of the materials furnished by this MS., without expressly quoting it.

2. A nearly complete series of correspondence between Professor Martyn and the late Dr. Pulteney, continued for thirty-six years.— The letters on the Professor's side, were obligingly communicated to the author, by Charles St. Barbe, Esq., of Lymington; those written by Dr. Pulteney, are preserved in the British Museum, among the MSS. of the late Sir Joseph Banks.

3. Correspondence between the Professor and several of his friends; particularly the Rev. George Ashby,—Mr. Charles Miller,— Sir Thomas Gery Cullum, Bart.,—Miss Hawkins,—and Sir James Edward Smith.

G. C. G.

Pertenhall, Bedfordshire,
Oct. 3, 1829.

Memoir

OF

JOHN MARTYN, F.R.S.,

BY HIS SON

THOMAS MARTYN, B.D., F.R.S., F.L.S.,

WITH ADDITIONS BY

GEORGE CORNELIUS GORHAM, B.D.

MEMOIR

OF

JOHN MARTYN, F.R.S.,

&c. &c.

THE subject of this memoir was born within a few months of the close of the seventeeth century. [He was descended from a family which, for several generations, had moved in the middle walks of life. The following particulars of his ancestry have been preserved; which, as connected with the birth of an individual who became so eminent in the history of Science, may not be altogether uninteresting to those who delight in Biographical researches.]

1. JOHN MARTYN, his ancestor of the sixth remove, is the first member of his family of whom there is any certain record. He was born about the beginning of the sixteenth century. He married *Margaret*, daughter and heiress of Humphrey *Ruiding*, of Droitwitch, Worcestershire. His son

2. GILBERT MARTYN, married *Katharine*, daughter of Sir George *Boteler*, of Stounbrooke. The next in descent,

3. JOHN MARTYN, fifth son of Gilbert, was born in 1558. He was parson of Swindon, [Gloucestershire?] in 1581. He died in 1626, aged 68 years. He married a widow lady, Mrs. *Mary Copley*, by whom he left a son,

4. THOMAS MARTYN, born in 1601. He was admitted Sizar of Emmanuel College, in Cambridge, 1617; and took his degree of B. A., in 1621; and of M. A., in 1625. He was Vicar of Little Houghton, in Northamptonshire. [His name occurs in the Register of that parish in 1641.] Having taken the covenant, he was expelled from his living at the Restoration; after which he came to London, and lived retired.[a] He married in 1627, *Sarah*,[b] third

---

[a] [MS. Note of Dr. Rawlinson, among the private papers of Mr. Martyn's family.]

[b] Sarah Proud's grandfather, Richard Proud, was, in 1561, parson of Bourton, on Dunsmore, Warwickshire. (See Dugdale's Warwickshire, p. 195.) Her grandmother, Rose Proud, was daughter of William Nottingham, Bailiff of Ipswich; whose father was Steward to the Lord Abbot of Bury-St.-Edmunds, and built the south porch of one of the churches there. This Rose Nottingham was persecuted in the time of Queen Mary, [in 1555, for piously encouraging a Martyr on his way to the stake,] but she escaped the flames, [by hiding herself in Ipswich.] (See Fox's Martyr's, pp. 1547, 1894, edit. 1610.) Sarah Proud's elder brother, Richard, was born 1595; was a scholar in Caius College, Cambridge; M. A., 1619; he was Rector of Thrandeston, Suffolk, in 1622, whence he was expelled in 1662; he died 1666, without issue, and was buried in Scole, *alias* Osmondeston Church, in Norfolk.

daughter of Richard *Proud*, citizen and goldsmith of London. Thomas Martyn died in 1692, aged 91 years. He was buried at Chigwell, in Essex, against the pulpit. [A portrait of him, taken in his 62d year, was not long since existing at Houghton Vicarage, painted on a pannel; of which a copy, in pencil, remains with his family.]

5. JOHN MARTYN, son of the preceding Thomas, was born in 1635. He married *Rhoda*, daughter of Thomas *Hodges*, D.D.*, by whom he had

6. THOMAS MARTYN, (father of the subject of this memoir.) He was born in 1622. He married *Ka-*

---

* THOMAS HODGES, D. D., was Vicar of Kensington, and one of the Assembly of Divines; after the Restoration he was Dean of Hereford, and Rector of St. Peter's, Cornhill. (See Newcourt's Repertorium, Vol. 1., p. 681, 526.—Also Wood's Athenæ Oxon., Vol. 2., p. 714.) He printed two sermous:—1. " The Growth and Spreading of Hæresie: set forth in a sermon preached before the Honourable House of Commons, on the 10th day of March; being the day of their public fast and humiliation for the growth of Hæresie. By THOMAS HODGES, Minister of God's word at Kensington. Published by order of the House of Commons. London, 1647." 4to.—2. " Sion's Hallelujah: set forth in a sermon preached before the Right Honourable House of Peers, in the Abbie Church of Westminster, on Thursday, June 28; being the day of Public Thanksgiving to Almighty God for his Majesties safe return. By THOMAS HODGES, Rector Ecclesiæ de Kensington. London. 1660." His second son was NATHANIEL HODGES, M. D., who staid in London, and attended patients unhurt, during the great plague. He published " Loimologia sive pestis nuperæ apud Populum Londinensem grassantis Narratio Historica. Lond. 1672, 8vo." Republished in English, by Dr. Quincy, 1720.

tharine Weedon, in 1698. He was a Hamburgh merchant, residing in Queen-street, in the city of London,
and was a man of strict integrity, and exemplary
piety. He died in 1743, aged 81 years nearly.

7. JOHN MARTYN, (the distinguished individual whose life we are about to record,) was born
at his father's house in Queen-street, September 12th,
1699. It was a subject of frequent exultation with
him, that providence had thrown him into a country,
and produced him at a period, so fertile in genius and
literary accomplishments : it was the golden age of
learning in Britain ; and to converse with those heroes
who adorned it, was no mean advantage or glory to
one who knew how to value it.

Scarcely was he a twelvemonth old ere he lost, in
his mother, the guardian of his tender infancy :—she
died Nov. 1, 1700; having been married only two
years and a half. His father, proposing to breed him
up to the profession of a merchant, sent him in due
time to a private school in the neighbourhood. Here,
by his own industry, rather than by any advantages
of instruction, he had made a very good proficiency in
school learning, when he was taken (in the autumn
of 1715, at sixteen years of age,) from his beloved
books, to engage in the business of a compting-house,
as his father's clerk. In this occupation most young
lads of his age would have found their minds sufficiently employed : but he, insatiate of knowledge,
after the labours of the day were over, stole most of

those hours which are usually given up to rest, and
dedicated them to the improvement of his understanding ; for several years seldom taking more than
four hours sleep.

Mr. Martyn, with good solid parts and unwearied
industry, was, however, no miracle of genius : he did
not begin with Vaillant at five years of age to be
acquainted with Botany.   His propensity to this
science, which was ever his favourite, appeared early,
namely in the summer of 1718, (June 30,) when he
had nearly attained his nineteenth year ; it was first
excited by an acquaintance with Mr. Wilmer.[a]   He
was further encouraged in his youthful taste for the
elegant science, which he was afterwards to advance
and adorn, by an acquaintance formed in August,
1719, with the celebrated Dr. Patrick Blair[b], whom he

[a] JOHN WILMER was an Apothecary in London. He afterwards
practised Physic at Chelsea, and became Reader in the Botanic
Garden there. From that place he retired to Westminster, and
died in January, 1769. [Some of his correspondence with Mr.
Martyn is preserved in the British Museum, among the papers
given by Professor Thomas Martyn to Sir Joseph Banks.]

[b] PATRICK BLAIR, M. D., F. R. S., was born in Scotland, [and
settled at Dundee as a surgeon.] He was a nonjuror, and was
imprisoned on suspicion of having been concerned in the Rebellion
of 1715. [In a letter, written in Newgate, to Sir Hans Sloane,
and dated Oct. 13, 1715, he declares that, he was, in no respect,
accessory to the late troubles ; but, happening to reside near the
parts in which the rebellion broke out, the gentry forced him to
accompany the army as a medical attendant. (MSS. Brit. Mus.
Sloane, 4038.) On July 8th, 1716, he wrote again to his friend
Sir Hans, informing him that his life was to terminate on the fol-

always acknowledged for his preceptor in Botany, and valued as the most intimate friend of his early life. [At the commencement of the following year, 1720,

lowing Friday. There is an incorrect story (recited in an amusing volume, entitled " Mems, Maxims, and Memoirs, by W. Wadd, Esq., F. L. S., 1824,") in which it is asserted, that " *Dr. Martyn,* Professor of Botany at Cambridge, supped with Dr. Blair in New- gate, the night previous to his expected execution." The anecdote adds, that Blair had been all along confident that he should be reprieved. Dr. Martyn said, " he sat pretty quietly till the clock struck *nine,* and then he got up and walked about the room ;—at *ten* he quickened his pace ;—and at *twelve,* no reprieve coming, he cried out, ' By my troth, this is carrying the jest too far !' The reprieve, however, came soon after, and in due time a pardon." The incident is a sufficiently probable one, but certainly is errone- ously connected with the name of *Mr. Martyn ;* who was then only seventeen years old, and was not acquainted with Dr. Blair till three years after. Mr. Martyn's subsequent intimacy with Dr. Blair may have given rise to this mistake ; but the person whose name should have been inserted in this story was the celebrated Botanist, Mr. Petiver, an Apothecary in Aldersgate Street, to whom Dr. Blair addressed a note, which still exists, (MSS. Sloane 4038,) written from Newgate, in the greatest agitation, soliciting his friend's company in the prison. His pardon seems to have been obtained by the influence of Sir Hans Sloane, and the exertions of the medical gentlemen in London. After his release, he did not return to Dundee ; and, having lost his business, was greatly distressed in his circumstances. In 1719, he wrote to Sir Hans Sloane, complaining that he was nearly ruined.] Having practised physic for a short time in London, he removed to Boston, in Lincolnshire, in April, 1720. [At this place he resumed his profes- sion under more favourable circumstances : he continued there till his death, which took place 1728. A great many of his letters

we find him pursuing his studies with ardour, notwithstanding the difficulties and interruptions with

are preserved in the British Museum ; viz., to Sir Hans Sloane, (MSS. Sloane, 4038, pp. 22—137,) from the year 1705 to 1728; to Mr. Petiver, (MSS. Sloane, 3321, pp. 32—77,) from 1712 to 1715,—and to Mr. John Martyn, (Sir J. Banks's MSS.) from 1720 to 1727.]   Dr. Blair was a warm admirer and defender of his countryman, Dr. Morison, [whose pretensions, as having given the earliest hints for a methodical arrangement of plants, he advocated with more nationality than judgment, against the claims set up for the illustrious Ray by his followers.]   He was also a great advocate for the circulation of the sap in vegetables.   He was a good Naturalist, and attended particularly to Botany and Ornithology.   Dr. Blair published—1. " *Botanick Essays, &c. &c.*" The first Essay treats of the structure of flowers ; the second, of the fruits ; the third, of the different methods of disposing plants ; the fourth, of their sexes, &c. ; the fifth, of their nourishment : the volume contains pp. 414, besides the preface.—2. " *Pharmaco-Botanologia: or, An Alphabetical and Classical Dissertation on all the British Indigenous and Garden Plants of the New London Dispensary, &c. &c.*"   London, 4to, 1723—1728.   This work was published in Decads, but was left incomplete, terminating with the seventh, at the letter H.—3. " *A new Table of Dispensatory Plants, distributed according to their principal virtues.*" Engraved on a folio copper-plate, and published at the desire of Dr. Mead. — 4. Several Papers in the *Philosophical Transactions ;* see Nos. 326, 327, 333, 358, 369, 364.   [Dr. Dale vexed him much, by fixing on him the contemptuous appellation, " *Proletarius scriptor iste !*" in a Medical Thesis, read in 1724, for his degree, at Leyden, and published—(See Dr. Blair's Letter to Mr. John Martyn, 9 May, 1824,) among the MSS. of Sir J. Banks, in the British Museum.—Linnæus, however, sensible of the merits of this zealous Botanist, has dedicated the genus BLÆRIA to his memory.]

which he had to contend, arising from his dull
engagements in the compting-house.   It is as useful
as it is interesting, to trace the earliest dawn of
genius ; however unimportant *in themselves* may be
the objects and investigations to which it first applies
itself.   The real lover of science will not, therefore,
consider the anecdote as insignificant which informs
him, that the young Naturalist was, at this time, in
the habit of rising at day-break, that he might improve himself in his favourite pursuit.   We trace him
in the " spring mornings" of 1720, sallying forth from
" his house, next the Rummer Tavern, in Queen-
Street, near Cheapside ," escaping for a few hours
from his detested desk and ledgers,—and hastening,
from the smoky atmosphere of the city, to the *then*
comparatvely rural neighbourhood of " St. George's
Fields," on " herbarising " excursions with his friend
Dr. Blair.   We might be inclined to smile at his
recording, on the 5th April, 1720, that he had then
found the vulgar herbs, " *Chelidonium minus* " and
" *Lamium rubrum*," were it not that these are the unaffected memoranda of a young student in Natural
History, who was *then* feeling his way to more important discoveries, and who only eight years afterwards, delighted the greatest Botanists of the age by

.   [Such is the endorsement of a letter addressed to him about
this time.]

  [Letter from Mr. Martyn to Mr. Wilmer, 5 April, 1720.]

  [*Ranunculus Ficaria*, Pile-wort Crowfoot.]

  [*Lamium purpureum*, Red Dead-Nettle.]

the first part of his splendid work,—the " *Historia Plantarum Rariorum.*"]

The sparks which had been thus kindled by his friend Mr. Wilmer, and fanned by his intimacy with Dr. Blair, were roused into flame by the countenance of the excellent and illustrious Sherard,* to

---

* The celebrated WILLIAM SHERARD, LL.D., F.R.S., [was born at Busby, in Leicestershire, 1659 ; and having been educated at Merchant Tailors' School, went to St. John's College, Oxford, in 1677, of which he became Fellow.   He travelled over many parts of England, and visited Jersey, for Botanical purposes.   He was appointed travelling tutor to Charles, second Viscount Townsend ; and afterwards to Lord Holland, who became Duke of Bedford in 1700.   He made two successive tours through Holland, France, Italy, &c., from whence he returned not much before the year 1700, as Sir James Smith thinks.   About 1702, he was sent as Consul to Smyrna ; in the neighbourhood of which, at Sedekio, he had a villa, where he began his great Herbarium : " Hasselquist visited the spot," says Sir James Smith, " with the devotion of a pilgrim, in 1749."]   In 1705, along with Antonio Picenini, he visited the seven churches of Asia : in 1709 and 1716, he transcribed the Monumenta Teïa ; he also caused the famous Sigean Inscription to be copied and sent to England, and the learned Chishull dedicates his account of it to him.   He returnd to England in 1718 ; and in 1721, he again visited France, Holland, and Italy.   As he was creeping on the Alps in search of plants, he narrowly escaped being shot by a peasant for a wolf. [—The same accident nearly befel Linnæus in Norway.   It is believed that Sherard again went to the continent in 1724, or 1725.]   Though he had acquired a considerable fortune during his stay in Asia, yet he lived with the greatest privacy in London ; wholly immersed in the study of Natural History ; except when he went to his brother James's seat and fine garden

whose acquaintance he had been introduced, Nov.
3d, 1719, and the knowledge of whom he reckoned
among the prime felicities of his life.' [The following
extracts from his correspondence will show how

at Eltham. He died, August 12th, 1728.—Dr. Sherard never
published any book *under his own name ;* but all the botanists of
his time acknowledge his assistance, and celebrate his praises ; as
Bobart, in his Preface to the last volume of the Historia Oxoniensis, and Ray, in the third volume of his Historia Plantarum.
He published the " *Schola Botanica,*" under the initials, S. W. A.,
(i. e. *Sherardus Wilhelmus Anglus,*) in 1689 ; with a Preface,
under the same signature, dated London, 1688 : this small
volume contains part of Tournefort's Botanical Lectures, or his
systematic List of Plants in the Paris Garden. He edited *Herman's Paradisus Batavus,* 1698, in the preface to which, dated
Geneva, 1697, he appears in his own name. Having purchased, in
1722, M. Vaillant's collection and papers, he assisted the learned
Boerhaave in the publication of *Vaillant's Botanicon Parisiense,*
1727, to which he prefixed an Epistle to Boerhaave. There are
several papers of his in the *Philosophical Transactions :* see
Nos. 274, 314, 367. The Third Edition of *Ray's Synopsis
Stirpium Britannicarum,* was published by Dillenius, in 1724,
under Dr. Sherard's inspection. The chief employment of his
retirement was his *Pinax,* or collection of names which had been
given by Botanical writers to plants. At his death, he bequeathed
the MS. of his *Pinax,* his library, and his valuable collection of
specimens, to the University of Oxford ; to which he also gave
£3,000 to provide a salary for a Professor of Botany. [Some
of] his MSS. were presented, in 1766, by Mr. Ellis, to the
Royal Society. '[The SHERARDIA, a humble British Plant, was
dedicated to his memory by Dillenius. There are a great many
of Sherard's Letters to Dr. Richardson, of North Bierly, published in Nichols's Illustr. of Lit. Anec., Vol. i., pp. 389—403.]

much, he valued the opportunities of improvement which his new acquaintance might afford, and how kindly the juvenile inquirer was encouraged by that distinguished Botanist.

[*April 5th*, 1720. *To Mr. Wilmer.* " Dr. Sherard has been so full of business with his brother's stoves and green-house, at Eltham, that I have not been able to see him since Christmas. . . . Dr. Blair goes next Monday to Boston."]

[*April 9th*, 1720. *To the same.* " On Thursday last, I had the happiness of above an hour's conversation with Dr. Sherard; who showed me some of the most beautiful figures of some species of *Lichen*, engraved upon copper-plates ; they are curiously described in all their different states : I hope they will be published in a little while. He informs me that the *Myosurus* (Mouse-tail,) grows plentifully at a place called Weston-Green, in the way to Eltham. . . . I have got a tin box made after the pattern of yours, &c. &c." . . .]

[*May 17th*, 1720. *To Dr. Patrick Blair.* " Though your staying in town would have been a very great advantage to me, I should have found a much greater want of you, if you had not been so kind as to recommend me to that worthy gentleman, Dr. Sherard. I waited on him this morning, &c. &c. . . . I herbarized, on Tuesday last, with the apothecaries : we made a pretty good collection, and Mr. Rand[a] showed

---

[a] [Mr. Rand was Lecturer and Demonstrator to the Company of

us that the *Lychnis* sylvestris rubello flore, was male
and female in diversis plantis."]

[*June 6th*, 1720. *To the same.* " I communicated
your letter and species of *Absynthium* to Dr. Sherard,
who took some of them down to his brother's house
at Eltham, in order to observe them as they grow. I
showed him your species of *Gramen;* it is the[b] *Gramen exile duriusculum, in muris et aridis proveniens,*
Raii Syn., edit. iii., p. 259."]

[*August,* 20*th,* 1720. *To the same.* " I herbarized
with the apothecaries this month. We dined at the
Green-Man, at Dulwich, and made a pretty good collection."]

[Dr. Blair was particularly struck by the amiable
character and ardent mind of his young friend; ·and,
(as appears by the following extracts from two letters,
dated from Boston, in May and in August, 1720,)
he saw from the very first, that Mr. Martyn would
soon distinguish himself in the walks of science :—

["Agreeable, sweet youth,

" Your truly pious and virtuous inclinations, do as
much engage me to keep a correspondence with you,

Apothecaries, in their Botanic garden. See some account of him
in Dr. Pulteney's Sketches of the progress of Botany, &c., vol. ii.,
p. 103.—Houstoun gave the name RANDIA, to a West Indian
shrub of the Pentandrous class, and it was retained by Linnæus;
but it is now removed to the Genus GARDENIA, (*G. Randia.*)—]

[a] [*Lychnis dioica a,* or Red Campion, of Eng. Flora.]
[b] [*Poa rigida,* or Hard Meadow-Grass, of Linnæus.]

as your other dispositions to those laudable studies to which I find your natural genius leads you, in which I am ever willing to encourage you, even at this distance. . . ." " I rejoice very much in your good acceptance by two such· eminent Botanists as Dr. Sherard and the good Mr. Rand. It is with pleasure that I behold you still going on, employing your leasure hours after such a manner; being in great hopes to see you one day making a considerable figure in the Botanic Chorus; to further you in which my small endeavours shall never be wanting. .

Your own,

PATRICK BLAIR."

—This promise was quite sincere; for Dr. Blair maintained an uninterrupted correspondence with Mr. Martyn, to the close of his life; and introduced him to the acquaintance, not only of Dr. Sherard, but also of Dr. Douglas, Dr. Massay, and Sir Hans Sloane, — connections which were of considerable value to him.]

Desirous, from the first, of communicating to mankind the fruit of his studies, he could not be long ere he designed something for the public eye. It was a good maxim, he used to say, whatsoever Persius might think of it,

" Scire tuum nihil est, nisi te scire hoc sciat alter."

" ——'Tis nothing worth that lies conceal'd,
  And Science is not Science till reveal'd."

Persius, Sat. i., Dryden's Trans.

Accordingly, with the opening of the year 1720, (January 11th,) he began a translation of Mr. Tournefort's excellent History of the plants about Paris; which he finished entirely on the 20th of August, following; although he deferred the publication of his volume till the year 1732, on account of the expected new edition by Vaillant. Already ambitious of rivalling this eminent French Botanist, he projected, in the month of March, of the same year, a Catalogue of the plants about London; in which he made a very considerable progress: but the world was deprived of this work through the envy and jealousy of some who preferred their own to the public advantage.

This year he began his botanical excursions, which he continued for a long time with unwearied diligence; not satisfied till he had collected together almost all the known plants of his native country: his *Hortus Siccus*, containing near 1400 specimens, is a sufficient testimony of it. Indeed, the great love which he possessed for his native soil, and the strong conviction which was ever upon his mind, that observations made upon plants, in their places of spontaneous growth, were the least liable to error, made him always pay the greatest attention to the natural productions of these kingdoms. It will be scarcely credible, in this age of ease and luxury, that almost all these excursions, some of them very extensive, were performed on foot. Nor did he confine himself wholly to the contemplation of vegetables;

the numerous insect tribe began now to attract a share of his attention. And when business, weather, or the season kept him at home, he corresponded diligently with his friends on these subjects, and sowed numberless plants in order to attend to the difference of their seed-leaves. [His friend Dr. Blair first introduced him to the notice of the public as an Author, by inserting a little piece of Mr. Martyn's in his own "Botanick Essays," (pp. 326—330,) which he published in 1720; viz. "*An Ode, formerly dedicated to Camerarius, translated into English,*" from the Latin: "*by* J. MARTYN, Φιλο-βοτανικος." The original is to be found in Camerarius's Epistle, " De Sexu Plantarum." If the translation have no particular merit, it at least shows the ardour of mind of the young student, whose own feelings we may well consider as appropriately described by the concluding stanza—

> Oh ! with what joy my eyes behold,
>   The wond'rous frame of nature's laws !
>   How my aspiring thoughts rejoice
> These myst'ries to unfold.]

The next year, (1721,) he framed an Introduction to the knowledge of Vegetables, (which he afterwards printed, in 1729,) and he was busied in making a complete collection of the fruits of all plants. He also became acquainted with the celebrated Dillenius[*],

---

[*] JOHN JAMES DILLENIUS, M. D., was born at Darmstadt, in 1681. He was brought over into England by Dr. Sherard, in August,

in conjunction with whom, and several others, he
instituted a meeting of Botanists in London, under
the name of 'The Botanical Society.'   They held
their meetings at first at the Rainbow Coffee-house,
in Watling-street, but afterwards in a private house,
every Saturday at six in the evening; and, in the
year following, formed themselves into a more regular
Society, with a President, a Secretary, and a body of
laws. Dr. Dillenius was their first President*, and Mr.

1721, to be his amanuensis, and after his death was Sherardian
Professor of Botany, in Oxford, till the year 1747, in which he
died.—He published, 1. " *Catalogus Plantarum circa Gissam
nascentium.* Franicof. 1719." 8vo.—2. " *Hortus Elthamensis.*
1732." folio.   In this work are contained 417 rare plants, with
their descriptions and synonymes ; adorned with very good and accurate figures, done by his own hand, from plants in Dr. James
Sherard's garden, at Eltham, in Kent. [This work did not sell,
and Dillenius cut up a part of it to hold his Hortus Siccus.]
—3. " *Historia Muscorum.* 1741." 4to.; containing the synonymes, history, and descriptions of above 600 Mosses ; with figures,
done also by himself.   The accuracy with which he has treated
this class of plants, (on account of their numbers and affinity
extremely difficult to distinguish,) and the great deficiency
of all former writers upon this subject, render Dillenius's work
very valuable, and have gained the author immortal honour.
—[4. He published, under the direction of Sherard, the third
edition of " *Raii Synopsis Stirpium Britannicarum.* 1724.'
8vo.—Linnæus named the Genus DILLENIA after him.]
   * [A Letter from Mr. Martyn to Sir Hans Sloane, dated Sept.
23d, 1725, exists in the British Museum, (MSS. Sloane, 4052,
p. 277.) in which he strongly presses Sir Hans to accept the
office of President of this Society.   The papers relating to the

Martyn their first Secretary. The other members
were, Samuel Horsman, M.D.; T. Richmond; Vincent Bacon*, F. R. S., a surgeon; J. Chandler;
John Wilmer, beforementioned, (p. 7.); Robert Fysher, M.B.; Samuel Latham; Thomas Dale, M. D.;
Philip Miller, the since celebrated Curator of the
Botanic Garden in Chelsea; Joseph Forsitt; J. Le-
therland, M. D.; Charles Deering[b], M. D.; Walter
Tullidelph[c]; Joseph Harris; and John Paine, an

Society were given by Professor Thomas Martyn to Sir Joseph
Banks.]

* VINCENT BACON, F. R. S., was a surgeon and apothecary; he
practised first in London, and afterwards at Grantham, in Lincolnshire. There is a paper of his in the Philosophical Transactions, No. 432,—" *The case of a man who was poisoned by
eating Monk's-Hood*," (*Aconitum Napellus.*)

[b] CHARLES DEERING, M.D., [was a native of Saxony.] He
came over to England in the train of a foreign Ambassador, [and
settled in London, as Dr. Pulteney thinks, about 1720.] He
removed to Nottingham, [in 1736, by the recommendation of Sir
Hans Sloane,] and practised physic there. [He died about
1749.] His works are, 1, " *Catalogus Stirpium*, &c., or *A Catalogue of plants naturally growing, and commonly cultivated
in divers parts of England, more especially about Nottingham.*
1738." 8vo.—2. " *A Historical Account of Nottingham, 1751.*"
4to.—[3. " *An Account of an Improved Method of treating
the Small Pox*, 1737," 8vo.—4. *Autobiographical Anecdotes* of
Himself, written in 1737; published in Nichols's Illustrations of
Anec. of 18th Century, Vol. I., pp. 211—220. The elegant
DEERINGIA CELOSIOIDES, (*Celosia baccata* of Willdenow,) a New
Holland Plant, so named after him, by Mr. Brown, commemorates
his love of Botany.]

[c] WALTER TULLIDELPH was amanuensis to Dr. Douglas. He

apothecary in London. The Society kept together
till the end of 1726.   Every member in his turn
was obliged to exhibit a certain number of plants,
to make observations upon their characters, and, to
set forth their various uses.   In 1721, Mr. Martyn
read before them a Course of Lectures, upon the
technical words of the science ; which was probably
the foundation of what he afterwards published upon
that subject.

This year, (1721,) and the following, (1722,) his
excursions in search of plants became more frequent,
in the neighbourhood of London, and the adjoining
counties of Middlesex, Surrey, Essex, and Kent.

In the summer of 1723, he pursued the study of
insects with more ardour: he entertained thoughts
of dissecting a variety of animals, in order to observe
their characteristic differences with accuracy, and to
enter upon the study of comparative anatomy : and
he laboured much at his designed Natural History of
the environs of London.

He still attended much to the *seed-leaves* of plants ;
upon which he had now conceived a design of forming a natural system ; and he made many observations upon the *sexes* of plants ; a subject at that
time little understood ; though it was well established
a few years after by the celebrated Linnæus.

afterwards settled in Antigua, as a planter; and in 1730, as a
practitioner in physic and surgery. [It is probable that he died
about 1739. Several of his letters, to Sir Hans Sloane, are preserved in the British Museum, MSS. Sloane, 4064.]

[We have already mentioned (p. 14,) that his attention had been directed, by Mr. Rand, to the diœcious character of the *Lychnis dioica;* one of those plants which then, and for some time after, puzzled even experienced Botanists; until repeated experiments, and particularly those of Linnæus, decisively established the doctrine of the sexes of vegetables. Mr. Martyn dug up a root of the female *Lychnis*, and planted it in a pot, that he might carefully observe its progress. The following extracts are from letters in which this subject, and that of the seed-leaves, were discussed with his friend Dr. Blair.]

From Mr. John Martyn to Dr. Blair.

Dec. 16, 1723.

Dear Sir,

I cannot tell which was greater, my surprise or pleasure, when I received your letter; in which you inform me that you are endeavouring after a method taken from the seed leaves. I have been actually engaged in the same design, and find so much encouragement, that I am resolved to push it on with vigour. I dare not say that I have seen every species of any one tribe, or one species of every tribe; but by what I have seen, I may form a conjecture that the *Leguminosæ* have their seed-leaves firm and carnous, and most of them not rising above ground: whether they may not upon this distinction form two tribes, remains to be inquired. The *Umbelliferæ* are, I believe, all narrow, long, and pointed. The *Stellatæ,*

oval, and in a manner cordate; which shows how
different this tribe is from the last; though they agree
in being *Gymnodispermous*. The *Tetrapetalæ Sili-
quosæ* are broad and cordate, but easily distinguish-
able from the *Stellatæ*. In short, I doubt, not but
I shall find the same concordance amongst the
Verticillatæ, Asperifoliæ, and perhaps *Multisiliquæ.*
I do not expect to find the *Apetalæ* agree together.
It is a class I am not at all fond of. But I should
not wonder to find *Lapathum, Atriplex, Bistorta, Per-
sicaria, Plantago,* &c., agree well in their seed-leaves;
and if they do, they may be classed together under
the name of *Spicatæ ;*—no new division, unless we
reckon Theophrastus a modern. The *Cucurbitaceæ*
may make a good class; and perhaps the *Malvaceæ*
too; and should I find the *Gerania* agree with them,
I should not wonder. *Lychnis, Caryophyllus,* and
other true caryophylleous flowers, may perhaps make
up a class under the title of *Coronariæ ;* a term used
by some old authors; as is also *Campanaciæ,* which
may make another. As for the *Monopetalæ* and *Pen-
tapetalæ Valculiferæ,* I have an aversion to them, as
well as to the *Apetalæ.* But enough of conjectures;
I do not, however, intend to build upon them. Experiments are what I want; and I do not doubt
of your assistance, as soon as the spring season shall
give us frequent opportunities of communicating our
mutual observations. . . .

JOHN MARTYN.

From the same to the same.

Jan. 14, 1723-4.

When I first entered upon the scheme of forming a method from the seed leaves, I thought their form would be sufficient to found distinctions upon : but on reading Cæsalpinus, (who gives more light into this doctrine than any of his successors,) I found that he had observed the situation of the point of the radicle in the seed, and had laid great stress upon it in classing of plants.

I agree with you in the observation of the *Cichoraceæ* having long, narrow, pointed seed-leaves, as well as the *Umbelliferæ*, but then the situation of the point of the radicle distinguishes them according to *Cæsalpinus.* This author puts the *Umbelliferæ* amongst those *quorum cor* (to use his own words) *exterius vergit :* and the compound flowers amongst those whose seeds have *cor in inferiore parte.* In this the *Scabiosa*, according to the same author, differs from compound flowers ; which makes me more inclinable to subscribe to the opinion of Dr. Dillenius, who separated it from them; because it has really *stamina* and *apices* like the simple flowers, and proper empalements to the little flowers. *Cæsalpinus* observes the same difference in the *Asperifoliæ* and *Verticillatæ*. . . . . .

JOHN MARTYN.

[Mr. Martyn's letters, relating to his experiments on

the *Lychnis dioica*, (p. 21) have not been preserved; but the nature of his observations may be conjectured from the replies of his correspondent. " Your objection," says Dr. Blair, (Feb. 20, 1723-4,) " concerning female flowers, sets me upon deeper thought. . . . . What you say of the *Lychnis*, if true, may destroy the very essence of the doctrine of the sexes of plants."]

[*From Dr. Blair to Mr. John Martyn.*

       Boston, Feb. 24, 1723-4.

[Good Mr. Martyn,

   I heartily agree with you that Natural Philosophy, especially its experimental part, might be brought up to as great a certainty as ever a proposition of Euclid, or any mathematical demonstration, were it not that our expression and imagination often disagree. . . . . . I was not a little delighted, when we came to communicate our thoughts concerning the *folia seminalia*, that you mentioned Cæsalpinus as giving the first hint of the circulation of the blood, upon which Harvey afterwards so handsomely enlarged; and this has frequently been seen in a great many discoveries made within these hundred years in Natural History, where the hints have been given by one, enlarged by the other, discovered by the third, and still greater improvements made by the fourth, &c. I have proposed, and indeed at Dr. Sherard's direction, Josephus De Arometariis for the first discoveries that the body of the seed consisted of the *folia seminalia* wrapt up; I doubt not he had the hint from Cæsalpinus, by what you write; but Dr. Grew was he who made the full discovery. I have shown what lively hints Senertus gave of the *sexes* of plants, and yet himself knew nothing of it. One might think the sexes of plants were known to Malpighi, by what you have quoted with relation to the dissection of a flower; and you see the foundation was laid by Dr. Grew, his cotemporary. How early did Mr. Ray take it; and yet did not think it worth his while to inculcate it with that vigour he might have done to spread it abroad. You see Tournefort is very particular in describing the several parts of a flower, and yet takes no notice of what Grew, Ray, and Camerarius had strenuously writ on the subject. . . . . To bring the point nearer home: what you have now observed to me concerning the *Lychnis*, did not, twenty years ago and upwards, escape the consideration of Mr. Bobart, at Oxford. You will find I have mentioned it in the Botanick Essays, as communicated to me by Dr. Sherard.* You have happily come to improve upon his

---

* [Dr. Sherard informed Dr. Blair of the following experiment, made about the year 1700.—Mr. Jacob Bobart, Overseer of the Physic Garden at Oxford, sowed some seeds of the *Lychnis sylvestris simplex* (*Lychnis dioica,*) "from a plant whose flowers had stamens without apices," (i. e. anthers,) but no plants sprang up. Hence Dr. Blair justly concluded that different sexes are necessary for the production of perfect seed. See Blair's Botanick Essays, p. 243. London, 1720.]

hint, and, I doubt not, to bring the discovery a greater
length. . . . . . I neither acknowledge nor reject a
female flower. If you will still have such, you will
find, in the MS.* in your hands, that I wrote to Mr.
Miller concerning his noted experiment of the Cab-
bages—'*one swallow makes no summer.*' I should have
looked upon that as merely accidental, had he not
wrote to me a second time that the same had been
confirmed by several other gardens. . . . . It is not
that species alone, of the *Lychnis rubello flore* you
mention, but even other species of *Lychnides*, that
has shown the same thing ; and indeed your experiment comes up to Mr. Bobart's as near as can
be ; but then, redundat quæstio,—Is this a Lusus
Naturæ or not ? This must depend upon the observation of the future season, &c. &c. . . . . Let the
Caryophylleous tribe be strictly examined by you and
all your other acquaintances. I shall not fail for my
part ; and if it amount to a discovery, I shall freely
own it to be *yours.* . . . My worthy friend, I heartily
thank you for your objection : it has been a mean to
wrest out what I did not before think of. I
write this the first time I put pen to paper in my
Green-house. . . . . .   Your own,

                                PATRICK BLAIR.]

[Mr. Martyn was, about this time, through Dr.

---

* [One of the Decads of the " Pharmaco-Botanologia," the whole
of which Mr. Martyn carried through the press.]

Blair's means, introduced to Sir Hans Sloane.   The
letter which brought him into acquaintance with that
patron of science, is worthy of notice, as showing
(what indeed has been observed above) how sensible
Dr. Blair was of the merits of his young friend, and
how clearly he foresaw his future celebrity.]

From Dr. Blair, to Sir Hans Sloane, Bart.

Boston, Jan. 1, 1723-4.

[Much honoured Sir,

. . . . I have desired, of a worthy young gentleman, Mr. John Martyn, that he would be so kind as to
deliver these to you, along with the sheets I design
for the public. . . . . Let me intreat you to signifye
your pleasure by the bearer.   It is with great delight
that I tell you that, though he exercises a quite different occupation, he has in a few years made such a
progress in Natural History, particularly Botany, that
he already exceeds many of his age, and, *if days and
years are continued to him, it is my opinion, few will
excell him in these laudable studys in his generation.* . . .

PATRICK BLAIR.]

[On 20th January, Dr. Blair writes to Mr. Martyn,
as follows :—" You may wait upon Sir Hans Sloane,
at six o'clock, at the Grecian Coffee House.   You will
find him very affable and free.   I hope my letter will
give you the opportunity to become acquainted with
him, if you are not already.   He will be ready to

receive my character of a friend, and you know what you deserve of me on that score."]

At the beginning of this year his life was endangered by an inflammatory fever.

[It appears, from the following letter, that he had at this time an intention of publishing some work— probably his collections for the Natural History of the environs of London.]

From Mr. John Martyn to Sir Hans Sloane, Bart.

Pancras Lane, Mar. 10, 1724.

[Sir,

I hope you will excuse the freedom I take in sending these papers to you. I should have waited on you myself, two or three days ago; but I have been sometime confined by a fever; and though I am now pretty well recovered, I am not yet in a condition fit to come so far. I do not doubt, Sir, but I shall have the favour of your recommendation and encouragement of this undertaking; in which I assure you I have no other design than to promote this useful part of learning, of which you have always been so eminent an encourager.

I am, Sir,

Your most obliged, humble Servant,

Jo. MARTYN.]

He had an offer of being introduced into the Royal Society, both this and the preceding year; but modestly declined it. [The following letters refer to this subject.]

From Dr. Blair to Mr. John Martyn.

Boston, 9 May, 1724.

[I am big with the expectation of the *Synopsis.** It seems, by the specimen you was so kind as to send me, it will put on a new countenance.   I am in good hopes by Sherard's, Boerhaave's, Vaillant's, and Dillenius's means, method will be brought to a great perfection in this age ; and if you live to see that number of years I have done, I rejoice at the thought of *your own* contributions to the advancement of these great improvements now in hand.   I remember some time ago, I wrote you concerning the *locus natalis* of several British plants ; you may communicate them to Dr. Dillenius if you think fit.

Now that you are so well acquainted with Sir Hans Sloane and Dr. Sherard, it is my opinion you ought to put up your desire of being admitted F. R. S. Both of them are sensible how much you deserve such an honour ; and if your modesty will not allow you to propose it to them yourself, I shall do it for you. . .

PATRICK BLAIR.]

From Mr. John Martyn to Dr. Blair.

May 10, 1724.

Dear Sir,

I am very much obliged to you for thinking I

---

* [The third edit. of Ray's " Synopsis Stirpium Britanicarum," which was then preparing by Dillenius, under the direction of Sherard, and was published in 1724.]

deserve the honour of being admitted into the Royal
Society.   Dr. Douglas was so kind as to offer to
introduce me last year; but I declined it, because it
is my opinion that they ought to be very cautious in
the admission of their members, and should take in
none but those who have given convincing proofs of
their learning and abilities. . . . .

JOHN MARTYN.

From Dr. Blair to Mr. John Martyn.

July 1, 1724.

[Good Mr. Martyn,
        I must again repeat my solicitations for your
admission into the Royal Society.   You are sensible
how low Natural History is amongst those by whom it
ought to be most encouraged.   The addition of Dr.
Sherard and Mr. Rand before my departure was very
requisite, and if *you* were entered into that illustrious
body, it would still make a greater figure as to that in
which you have rendered yourself so far known.   Let
not your modesty prevail too far with you. . . . .

PATRICK BLAIR.]

He was now (1725) so far advanced in his favourite
pursuit, that he ventured to instruct others, by giving
a course of Lectures in Botany at London*.   He also
gave an explanation of the technical words used in

* See the Preface to Dr. Blair's 4th Decad of his *Pharmaco-
Botanologia*, p. 2.

that science, to Mr. Nathan Bailey, in order to be inserted in the Dictionary which he was then about to publish. He continued his excursions in search of plants; and was much busied in forming, with the assistance of Dr. Blair, a collection of birds, in order to make observations and improvements in that branch of Natural History.

Besides the repetition of his home circuits, in the months of June and July of this year, he travelled by Bath and Bristol into Wales; returning by Hereford, Worcester, and Oxford, to London. This journey furnished him with a great variety of new plants, such as are not to be found in the neighbourhood of London. Amidst his excursions, however, he did not forget to extend his observations on seed-leaves, and on the sexes of vegetables.

In the succeeding year (1726) he read a second course of Lectures in Botany at London. He now published his first work; his Tables of Officinal Plants, under the title of " *Tabulæ Synopticæ Plantarum Officinalium ad methodum Raïanam dispositæ. Londini,* 1726 ;" folio, pp. iv. and 20 ; dedicated to Sir Hans Sloane, Bart. Twice in the course of this year he visited the Isle of Shepey, so well known among Naturalists for its curious productions, both vegetable and fossil.

Some circumstances which occurred about this time at Cambridge, gave him an. opportunity of employing his talents in that University. Richard Bradley, F. R. S., the well-known author of various treatises in Natural History, was at this time Professor of Botany there. He was chosen into the office Nov. 10, 1724, by means of a pretended verbal recommendation from Dr. Sherard to Dr. Bentley, and pompous assurances that he would procure the University a public Botanic Garden by his own private purse and personal interest. The vanity of his promises was now seen, and his total ignorance of the learned languages known: so that, as the Professor neglected* to read lectures himself, the University

---

* Mr. Bradley, however, read a Course of Lectures on the Materia Medica, in 1729. (See the Grub-Street Journal, No. 11.) In 1731, he was grown so scandalous, that it was in agitation to turn him out of his Professorship ; and he died in the latter end of 1732.—It may seem strange to assert that, the Translator of Xenophon's Œconomics did not understand Greek: it is, however, true. Mr. Bradley's being then a popular name, he was paid by the booksellers for permitting them to insert it in the title. He might, however, have made *this* translation without much knowledge of the Greek language ; for, upon examination, it turns out only to be an old translation modernized ! [He published the Lectures above alluded to, (a mean performance,) under the following title : " *A Course of Lectures upon the Materia Medica, read in the Physick Schools at Cambridge, upon the Collections of Doctor Atterbrook, and Signior Vigani, deposited in Catharine Hall, and Queen's College. By R. BRADLEY, F. R. S., and Professor of Botany in the University of Cambridge.* London. 1730." — Dr. Pulteney has thus drawn his character :—" The industry and talents of Bradley were not mean ; and, though unadorned by learning, were sufficient to have secured to him that reputable degree of respect from posterity, which it will ever .justly withhold from him who fails to recommend such qualifica-

made no difficulty to permit another person to do it.
Mr. Martyn was recommended by Dr. Sherard and Sir
Hans Sloane as a proper person to execute the office.
Accordingly, the next year, (1727,) in the Anatomy
Schools, he gave the first Course that ever had been
read there in that Science, with a view to restore this
study on the spot which should seem most adapted to
its growth, as having nourished the most eminent of
of all our English Naturalists, the excellent Mr. Ray.
[Dr. Sherard, in a letter to Dr. Richardson, of North
Bierley, dated London, 14 March, 1726-7, thus adverts to this subject :—" Mr. John Martyn, who gave
a College of Botany here last summer to several
young gentlemen, goes next month to Cambridge,
whither he is invited by above twenty scholars. He
carries on, at the same time, his College here ; spending April, May, and part of June, there; the rest of
June and July here; August at Cambridge; and
finishes here in September."[a]]

March 30, 1727, Mr. Martyn was admitted a Member of the Royal Society. His modesty prevented
(as we have seen) his more early admission; but
there was one circumstance on which he used to congratulate himself—that he was proposed, *though the*

tions by integrity, and propriety of conduct. In these, unhappily,
Mr. Bradley was deficient." Pulteney's Historical and Biographical Sketches, Vol. II., p. 129. His works amounted to 2 Vols.
folio, 4 in 4to., and nearly 20 in 8vo.; of which an account is given
in Nichols's Literary Anec. Vol. I., pp. 446—451.]
  [a] Nichols's Illustr. of Lit. Anec., Vol. I., p. 401.

D

last, under the Presidentship of the illustrious Newton; who died on the 20th of March, before Mr.
Martyn's admission.   As he was unwilling to be
admitted of this learned body whilst he deemed
himself unworthy of that honour; so, after he *was*
admitted, he thought himself under an obligation to
be as serviceable to it as possible.   Accordingly he
was often put upon Committees for various purposes;
particularly, in the year 1729, upon one for inspect-
ing and regulating the Library and Museum; in
which he took the lead, and drew up the report, in
1731.   For these services he received the thanks of
the Society, and had his bond for the annual payment
of fifty-two shillings cancelled, by an order of Council,
as an acknowledgment of his merits.   He also en-
riched the Philosophical Transactions with many
valuable papers; which will be noticed under their
respective dates.

This year, (March 31,) he printed, for the use of his
pupils, at Cambridge, " *Methodus Plantarum circa
Cantabrigiam nascentium. Londini,* 1727 ;" 12mo. pp.
viii. and 132.   It is Mr. Ray's alphabetical Catalogue
reduced to the *order* of his system; with excellent
generic characters, taken from the " *Methodus emen-
data et aucta,*" of the same writer, from Vaillant, Dillenius, Scheuchzer and others,—improved and corrected by his own observations.   [In the Preface it is
stated, that this book, though printed, was *not published;* being intended merely for his pupils, and
judged unfit for the public eye; as having been writ-

ten in a few days, and rapidly carried through the press.] About two years after, he designed a second edition, much improved and enlarged; containing also about 150 plants more than Mr. Ray had found in Cambridgeshire. A sheet and a half of this was printed, but it was never carried any further;—it contains the Submarine plants, Funguses, Mosses, Cappillaries, Apetalous, and Jeliferous plants; to the number of 154; most of them not to be found in Mr. Ray's catalogue. [In fact, he spared no pains to bring the science of Botany into repute in the University; " perambulating the country with his scholars, and showing them the Cambridgeshire plants where Mr. Ray had described them to grow." He was not less diligent in preparing. his Metropolitan Flora, Dr, Sherard, in a letter to Dr. Richardson, dated Aug. 5, 1727, writes—" Mr. Martyn, that has given a College of Botany this summer at Cambridge, is about to print ' a Catalogue of Plants within a day's walk from London.' "[b]]

The following year, 1728, he was busied in settling the natural classes of plants; a subject on which he had bestowed much pains and attention. He followed Mr. Ray in the distribution of his classes; he differed, however, from him in some things; particularly in uniting the trees and herbs; and certainly much improved the system of our learned countryman.

[a] [See " A Short Account of the late donation of a Botanic Garden to the University of Cambridge," 4to. 1763.]
[b] [Nichols's Illustr. of Lit. Anec., Vol. I., p. 402.]

[From the year 1720 to 1728, he gave very considerable assistance to Dr. Blair, by correcting the press for his works, conferring with scientific persons in the Metropolis, and by an uninterrupted correspondence with his friend, whom he lost about this time.]

He now published the first Decad of rare plants, under the title, " Joannis Martyn *Historia plantarum rariorum.—Juvat integros accedere fontes,Atque haurire, juvatque novos decerpere flores.* Lucret.—*Londini, ex typographia Richardi Reily,* 1728." Imperial folio. [The work contains pp. iv. and 52, with fifty plates.]   It was dedicated to the President, Council, and Fellows of the Royal Society.   The design of this sumptuous work was, to figure such curious plants as had never been figured before, of their natural size, and in their proper colours ; to give descriptions of them, and to add their culture and use.   The paintings were executed by Van Huysum, and the engravings, by Kirkall ; they are mezzotinto, and printed in their proper colours.   Four Decads more were published between this time and 1737, when the fifth and last came out ; each of them, as the name implies, containing ten plants.   [Among many other rarities, these figures exhibit several *Gerania,* the *Milleria, Martynia, Gronovia,* and *Turneria.*]   The work stopped at the fifth Decad, on account of the immense expense attending the plates.   John Daniel Meyer, a painter, at Nurimberg, republished the first Decad in 1752.

At this time he applied himself sedulously to the

practise of physic.   In the year 1729, he entertained
a design of reading a Course of Lectures in Botany at
Oxford; but, for what reason I know not, it proved
abortive.   This winter he hoped to finish his Catalogue of the plants about London, which has been
mentioned before (p. 28) : and he published " *The first
Lecture of a Course of Botany ; being an Introduction to
the rest.* By John Martyn, F. R. S. *London,* 1729,"
8vo., pp. viii. and 23, with fourteen plates. Dedicated to
Dr. Christopher Green, Regius Professor of Physic in
the University of Cambridge.   It is an explanation of
the technical terms made use of in Botany. This year
he also published his *Second Decad of rare plants.*

In the Philosophical Transactions of this year will
be found his first communication to the Royal Society.
" *An Account of some observations relating to Natural
History, made in a journey to the Peak in Derbyshire :*
by Mr. J. Martyn, F. R. S." (Phil. Trans. for 1729,
No. 407, Vol. XXXVI., pp. 22—32.)   [This paper
contains an account of some rare plants found at the
Peak, and in the journey thither, " not taken notice
of by the Bishop of London, (Gibson,) in his edition of
Camden."   Mr. Martyn here takes occasion to separate the *Lactuca sylvestris murorum flore luteo,*
of Bauhin and Ray, from that genus, and gives it the
name of *Scariola :* Linnæus justifies the distinction,
but calls it *Prenanthes :*—the plant is the *Prenanthes
muralis,* or Ivy-leaved Wall-Lettuce, of Dr. Smith's
English Flora.]

The year following, (1730,) he was engaged, in con-

junction with Dr. Russell, in a design of republishing
Roberti Stephani Thesaurus Linguæ Latinæ ; but either
because the proposal did not meet with due encouragement, or for some other reason unknown, the
design was dropped. He was also concerned with
the same learned gentleman and others, in a weekly
paper, entitled " *The Grub-Street Journal,*" the principal intention of which was to ridicule bad authors
and their works. Mr. Martyn wrote the introductory
paper under the title of *Bavius,* which was the character he preserved throughout this work; to which
the greatest wits of the time did not disdain sometimes to contribute. The best papers were afterwards
selected and printed in two volumes, 12mo., in the
year 1737, under the title of " *Memoirs of the Society
of Grub-Street.*" The papers which were written by
Mr. Martyn are distinguished by the signature B.
Dr. Russell took the title of Mævius, and his papers
are signed M. The Grub-street Journal had a large
sale, and was kept up till the end of the year 1737.
There was an attempt made to revive it, at the beginning of the year following, under the title of " *The
Literary Courier of Grub-Street ;*" but, probably, without much success, for it was soon dropped.

May 26th, 1730, he was admitted of Emmanuel
College, in Cambridge, and kept five terms, with an
intention to have proceeded regularly with the de-
grees in Physic; but marriage and the necessity of
attending to his profession prevented him from finishing his design. He read a course of Lectures this

year both in Botany and the Materia Medica, in the University; on the latter subject also in London. The pursuit of this study led him to a more intimate acquaintance with Chemistry, in which he was now much engaged in making experiments.

Mr. Martyn's second communication to the Royal Society appears in the Philosophical Transactions for this year. *" Memoirs communicated by M. Garcin to M. St. Hyacinthe, F.R.S., containing a description of a new family of plants called* Oxyoides [with a remark by the Translator]; *some remarks on the family of plants called* Musa; *and a description of the* Hirundinella marina, *or* Sea Leach. *Translated from the French, by Mr.* JOHN MARTYN, F.R.S." (Phil. Trans. for 1730, No. 415, Vol. XXXVI., pp. 377—394.) [This paper is illustrated by three copper-plates; 1. of the *Oxyoides Javanica,* and *Oxyoides Malabarica ;* 2. of the Seed-Vessel of the Genus *Musa;* and 3. of the *Hirundinella marina.* Mr. Martyn remarks that, M. Garcin was mistaken in supposing that the *Oxyoides Javanica* was a new species; having long before been described by Acosta, under the name of *Herba viva.* He says it was the first *sensitive* plant known in Europe; and points out that it was altogether a different plant from those now brought from America, and cultivated in our gardens under that name. The *Oxyoides Javanica,* and *O. Malabarica,* are the same species, and are now called *Oxalis sensitiva,* or *Sensitive Wood-Sorrel.* It shuts at sunset, and is quite as sensitive as the *Mimosa.* It does not appear to be

now known in English Gardens. It is figured in
Zannoni, Historia Botanica, t. 61.; Rhede's Hortus
Malabaricus, Part IX., t. 19; and Rumphius, Herbarium Amboinense, Part V., t. 104, fig. 2.—Jacquin's
beautiful Monagraph of the *Oxalis*, t. 78., f. 4., only
represents the curious stamens and pistils of *O. sensitiva*, highly magnified.]

He was this year an unsuccessful candidate for the
post of Secretary to the Royal Society, vacant by
the death of Dr. Rutty. His friend, Mr. Philip Miller, gave him the first intelligence of this opening,
in the following letter :—

From Mr. Philip Miller to Mr. John Martyn.

Chelsey, June 6, 1730.

Sir,
    Dr. Rutty is in a very bad way, and is thought,
by those who attend him, very near death ; so that, if
you intend to stand for his place, I think it absolutely
necessary for you to be in town; but this I submit to
your own choice; I have had an opportunity of talk-
ing with several members of the Royal Society, and
find your interest with them very considerable. Mr.
Eames told me, he thought you the most proper
person, as did Mr. Hales and some others. As to Dr.
Wiggan, though he makes great interest, and is es-
poused by Dr. Mead, who engaged Mr. Hales and
many others to join their votes for him, yet he cannot
stand, for he is not admitted a member yet, nor do I
believe he will be until the election is over; for there

has been no meeting since he was proposed, and I understand there will not be a council called, because he should not have an opportunity to stand; so that the interest of Sir Hans will be greatly divided, and probably a greater opportunity for a third person to carry it away. You may depend on it there will not be any opponent but Dr. Mortimer. As for Dr. Nichols, he has very little interest that I can find; and the mathematicians all oppose Dr. Pemberton, by which means if you can engage them for you, I think there will be no danger.

PHILIP MILLER.

[As soon as the death of the Secretary (which took place on the 10th of June) was announced, Mr. Martyn addressed a letter to Sir Hans Sloane, soliciting his influence :—

[*From Mr. John Martyn to Sir Hans Sloane, Bart.*

Emmanuel College, June 13, 1730.
[Sir,

I am informed that Dr. Rutty is dead, and that several of my friends propose that I should succeed him. I am far from desiring the place, if any gentleman better qualified will accept of it; but, if not, I should be glad to serve the Society in a place which may be of some little advantage to me; having hitherto been always ready to serve them without any view of profit. I have not made any personal applications, because I apprehend it is not usual. But I thought myself under the necessity of communicating

my intentions to you, whose favour and encouragement have been already of so much service to me,
least you should think I did not desire this place, on
account of the expectations of other preferments at
Cambridge.   You know how precarious Dr. Woodward's Professorship is, though I have followed your
advice in becoming a member of the University,
which will wipe off Mr. Windsor's objection.   The
Physick Garden will probably be some years before it
is established ; and *expectation* will not bring me food
and raiment, as I am already too sensible.—If you
have any commands here, be pleased to lay them on,

   Your most obliged
    Humble Servant,

    John Martyn

This application was not successful.] — His opponent, Dr. Mortimer, was related to the President;
and the interest of Sir Hans Sloane, united with the
whole power of the Court, was too prevalent for the
literary part of the Society. Mr. Martyn was peculiarly fitted for this post, by his extensive learning,
and his acquaintance with the modern languages : but
the countenance and support of Halley, Jurin, and
the rest of those who were the great ornaments of this
Society, made him full amends for his ill success.
 He had now lived and practised physic in Great
St. Helen's, Bishopsgate-street, during three years ;

but at the close of this year, finding the air of London
disagree with his constitution, on account of an asth-
matic complaint with which he was troubled, he em-
braced a favourable opportunity which now offered of
removing to Chelsea; where from this time he practised physic with tolerable success, and great reputation, for above twenty years; till increasing infirmities
forced him into retirement.

His friend, Mr. Houstoun, who was now on his
travels in South America and the islands of the West
Indies, in search of plants, gave the name MARTYNIA*
to a new and elegant genus of the Didynamous

* [MARTYNIA, belongs to the class and order *Didynamia An-
giospermia* of Linnæus,—and to the natural order *Bignoniæ* of
Jussieu, or *Pedalinæ* of Brown.

Generic character.—*Calyx*, 5-cleft; *Corolla*, ringent: *Stamina*, 5, the fifth being imperfect; *Stigma*, two-cleft; *Capsule*,
woody, with a hooked beak, 4-celled, 2-valved.

Specific characters :—

1. *M. diandra*, or *Two-stamened Martynia*. Stem, branched,
hairy, two feet high : leaves, opposite, heart-shaped, toothed,
hairy, the hairs terminated by clammy globules; flowers, two-
stamened, white tinged with purple; border, pale red with a shining purple spot on each segment; throat and middle segment,
spotted with yellow, and streaked with yellow lines resembling the
stamens and style with the two-cleft stigma: Miller observes, that
the blossom greatly resembles the *Fox-glove*. Found at Vera
Cruz by Mr. Houstoun, and introduced into England in 1731.
(Figured, badly, in Ehret, Plantæ rariores, t. 1. fig. 1.; well, in
Martyn's Historia Pl. rar. p. 42.; best, in Andrews's Repository,
t. 575.) Annual.—The most showy of its tribe.

2. *M. Craniolaria*, or *White-flowered Martynia*.—Stem,
branched, hairy : leaves, opposite, deeply 5-lobed, toothed: calyx,

class, which he had just discovered. [Five species of this genus are now known : that which was dedi-

an ovate sheath, split longitudinally in front: flowers, white, very long in the tube. Found at Carthagena, and introduced into England about 1733, by Mr. Houstoun. (Figured, indifferently, in Ehret, Pl. rar. t. 1. fig. 2. ; a detached flower and leaf well engraved in Jacquin, Stirpes Americanæ selectæ, t. 110.) Annual.

3. *M. proboscidea*, or *Horn-capsuled Martynia.*—Stem, branched : leaves, opposite and alternate, ovate-heart-shaped, very entire: flowers, tube yellowish-white, merging towards the border into a pale violet or lilac ; throat, exquisitely streaked with violet lines, and freckled with violet and saffron dots ; border, waved, wide, pale violet, middle segment of the lower lip largest, roundish, delicately streaked at the base with sulphur-coloured lines continued into the throat, resembling the stamens and style with the two-cleft stigma ; beak of the Capsule very long, and incurvated something like the *proboscis* of an elephant, whence the specific name. Found in Louisiana, about the Mississipi. Cultivated in England, in 1738, by Robert James, Lord Petre. (Figured by Miller, Figures of plants, fig. 286, tolerable ;— Schmidel, Icon., p. 49, t. 12, 13, excellent for the *drawing* of the flowers in various positions, but the *colouring* bad ;—Gærtner, de fruct. et semin. pl., Vol. 2, p. 131, t. 110, a figure of the seed-vessel only ;—Kretzschmar, in a Monograph on this plant, entitled " Beschreibung der Martyniæ annuæ villosæ, Friedrichstadt, 1764," 4to., pp. 26, with two copper-plates ;—and Curtis, Bot. Mag., t. 1056, but very badly as regards the colouring.) Annual. The most elegant of its family. This species has been appropriately selected as a vignette to the portrait of Professor Thomas Martyn, engraved in 1799, by Vendramini, from a painting by Russel.

4. *M. longiflora*, or *Long-flowered Martynia.* Stem, simple ; leaves, roundish, waved ; flowers, white ; tube of the corolla, long, contracted at the middle, swelling at the base on the back

cated to Mr. Martyn is the *Martynia diandra,* or *Two-stamened Martynia,* a native of Vera Cruz, and is perhaps the most showy of its family. It is figured by Mr. Martyn at the 42d page of his " Historia plantarum rariorum ; " where he thus modestly speaks of the honour which had been attached to his name : —" Plantæ huic, non minus eleganti quam novæ, hoc nomen imponere dilectissimo Houstouno placuit, ne amici sui periturum nomen omnino ignorarent posteri ; "—" *on this plant my dearest Houstoun has thought fit to bestow the name of his friend, that it might not sink into oblivion, and be altogether unknown to posterity.*" [The following letter notices this compliment :—

Found at the Cape of Good Hope. Introduced into England in 1781, by the Countess of Strathmore. (Figured by Meerburg, Afbeeldingen, &c., t. 7., a wretched representation as regards the colouring.) Annual.

5. *M. lutea,* or *Yellow Martynia.* Stem, hairy, with glands ; leaves, heart-shaped, round, toothed, hairy ; beak much longer than the capsule ; flowers, a rich golden yellow ; the throat and margin, freckled and stained with red crimson dots and streaks. Received from the Brazils, by the Hon. and Rev. W. Herbert, in 1824. (Figured in the Botanical Register, t. 934.) Annual.

All these tender annuals require the aid of a bark stove ; and as they are apt not to ripen their seeds, the plants are easily lost. Only the 3d and 5th species are now to be met with in English gardens. " Cultivators (says Sir J. E. Smith, in a letter to the editor of this volume, Aug. 5, 1827,) seldom keep up their collections of curious *annuals,* though the beauty of the MARTYNIÆ might have procured some favour for them."]

From Mr. John Martyn to Mr. Houstoun.

Chelsey, May 22, 1731.

Dear Sir,

The only letter I have had the favour of receiving from you, was dated December 14th, [1730]. But I have been favoured with some of your specimens by Mr. Miller. The last collection indeed has been almost spoiled by the sea water; and what few were saved, are mostly sunk a second time in the abyss of Sir Hans's collection, where I never expect to see, or hear of them more.

I am obliged to you for the remembrance of me, with some others of your friends, in the names you have ascribed to your new Genera.

I agree with you in your opinion that no more characters should be given of a Genus, than are sufficient to distinguish it from all others. This would be rendered more easy, had I time and abilities to finish those general tables which I have begun.

I just saw your account of the Cochineal, and must let you know with some concern, that, as it differed in some material circumstances from that of Sir Hans, the public will never reap the benefit of it; so far does some men's ambition go beyond an undissembled love of truth. For the future, if you have any thing which you would have communicated to the Royal Society, or the public, I would advise you to think of conveying it by some other channel. You have done enough already to stir up the jealousy of any man

who has been before you in the West Indies ; and, if
I am not grievously mistaken, you will soon make it
difficult for any one to have success who comes after
you. I wish those who are able were willing, or else
that those who are willing were more able to serve
you. If good wishes, and the highest esteem of your
diligence and capacity could be of any service to you,
you have them from,

Dear Sir, Your's, &c.
JOHN MARTYN.

This year, (1731,) he entertained sanguine hopes
that a Botanic Garden would have been founded at
Cambridge, by Mr. Brownell, of Willingham, a gentleman possessed of a handsome fortune, and a great
lover of the science. But though there passed many
conferences upon the subject between him and the
Vice-Chancellor, Dr. Mawson, then Master of Bennet
College, afterwards Bishop of Ely, and Dr. Savage,
the Master of Emmanuel College ; though the ground
was actually pitched upon, and Mr. Miller was called
in to deliver his opinion concerning the most proper
manner of carrying the design into execution ; yet it
dropped, and Mr. Brownell's estate was diverted into
another channel. He did not forget this year to contribute his endeavours to promote the cause of learning : he was much busied in forming Generical Tables
of plants ; published his *Third Decad of rare plants ;*
and was engaged in putting together Churchill's Collection of Voyages and Travels.

These were quickly followed, the succeeding year, (1732), by his long since finished work, the translation of Tournefort's history of the plants growing about Paris. It came out under the title of " *Tournefort's History of plants growing about Paris: with their uses in physic; and a mechanical account of the operation of medicines; translated into English, with many additions. And accommodated to the plants growing in Great Britain. By* JOHN MARTYN, F.R.S. *In two volumes. London,* 1732." 8vo. It is dedicated to Lord Petre. The translator reduced Tournefort's six alphabets into one; and extracted all the useful observations both from the edition which came out by the united care of Sherard and Boerhaave, and from that which was published by M. Bernard de Jussieu. He added also the English names, and the places where the plants grow in England. He disposed the Mosses according to Dr. Dillenius's method; and the Mushrooms and Capillaries after a new method of his own.

August the 20th, 1732, he married EULALIA, youngest daughter of JOHN KING,[*] D. D., Rector of Chelsea,

---

[*] JOHN KING, D. D., was born at St. Columb, in Cornwall, May 1, 1652. [His first wife was Anne, youngest daughter of William Durham, by Lætitia, grand-daughter of Sir Francis Knollys, first cousin to Queen Elizabeth, and treasurer of her houshold: he had no issue by her. His second wife (1690) was Elizabeth, daughter of Joseph Aris, Esq. of Adstone, Northamptonshire, widow of the Rev. John Eston: by this marriage] he became patron of the Church of Pertenhall, in Bedfordshire, and was in-

and Prebendary of York: she was born at Chelsea, July

stituted Rector there, in the year 1690, but in 1694 he removed,
by exchange, to Chelsea.   A great intimacy subsisted between
him and Sir William Dawes, Archbishop of York, who gave him
the Prebend of Wighton, in the Cathedral Church of York, in the
year 1718.   Though educated at Exeter College, in Oxford, he
took the degree of D. D., in 1698, at Catharine Hall, in Cambridge, where the Archbishop was Master.   He died May 30th,
1732, aged 80, and was buried at Pertenhall.   He published,—1.
" A Sermon preached at the Funeral of Sir Willoughby Cham-
berlain, Knt., who died at his house at Chelsea, Dec. 6, and was
interred at the Parish Church of St. James, Garlick Hith, London,
Dec. 12, 1697 : By JOHN KING, Rector of Chelsea, near London.
Printed for Thomas Bennet, at the Half-Moon, in St. Paul's
Church-yard, 1697." 4to.—2. " The Divine Favour the best
Alliance: or Repentance the safest sanctuary in times of Danger.   A Sermon, preached at the Parish Church of Chelsea,
near London, on Friday, the 19th of December, 1701.   Being
the fast-day appointed by his Majesty's proclamation: By JOHN
KING, D. D., Rector of Chelsea.   London: Printed for Thomas
Speed ; over against Jonathan's Coffee House, in Exchange Alley,
in Cornhill, 1702." 4to.— 3. " Animadversions on a Pamphlet
intituled, A Letter of Advice to the Churches of the Noncon-
formists of the English Nation ; endeavouring their satisfaction
in that point, *Who are the true Church of England?* By JOHN
KING, D. D., Rector of Chelsea, near London.   The Second Edition.   London: Printed for Thomas Speed, 1702." 4to.—4. The
case of John Atherton, Bishop of Waterford, in Ireland: fairly
represented.   Against a late partial Edition of Dr. Barnard's
Relation and Sermon at his Funeral.   With some remarks on
the Title Page, the additional Preface, and a Scandalous and
Immodest Advertisement, lately set forth in a Publick Paper ;
whereby the said Bishop's Case is misrepresented, and the Epis-

E

22, 1703: by her he had three sons and five daugh-

copal Order aspersed. With a brief account of a Conspiracy
against the life of Mr. Robert Hawkins, Minister of Chilton,
Bucks, Tryed before Sir Matthew Hale. Also of the plot of
Robert Young, and Stephen Blackhead, against the Bishop of
Rochester. London: Printed for Luke Stokoe, at the Golden
Key and Bible, near Charing Cross, and sold by J. Morphew,
near Stationers' Hall. 1716." 8vo. In the Appendix are two
Anonymous Letters; but it appears, by interlineations in Dr.
King's own hand, that the first was from Dr. Thomas Mill,
Bishop of Waterford; and that the second was to that Bishop
from the Rev. Mr. Alcock, Chancellor of Waterford.—5. To-
lando Pseudologo-mastix, or a Currycomb for a lying coxcomb.
Being an answer to a late piece of Mr. Toland's called Hypatia,
&c., London: printed for Luke Stokoe, and sold by T. Bicker-
ton, at the Crown, in Pater-noster Row. 1721." 8vo. [There
exists, in the British Museum, (MSS. Sloane, 4455,) a small
4to. Common Place Book, in the hand writing of Dr. King.
The contents are chiefly of a very trivial kind, but there are a
few pages of curious remarks, on Herne's edition of Roper's Life
of Sir Thomas More, and criticisms on the falsely supposed site
of his house, which Dr. King ascertained to be what was afterwards *Beaufort House.*] His eldest son, JOHN KING, was
born August 5th, 1696. From Eton School he was sent to
King's College, in Cambridge, where he became Fellow. He afterwards settled at Stamford, where he practised physic with great
reputation, but was cut off by a fever, Oct. 12, 1728. He
published, " Euripidis Hecuba, Orestes, et Phœnissæ; &c., &c.
Cantab. 1726." 8vo. By Lucy, daughter of Thomas Morice,
Esq., he had one son, JOHN KING, Rector of Pertenhall, in
Bedfordshire, from 1752 to 1800, who died, Oct. 6, 1812. [The
Poet, Cowper, corresponded with the amiable wife of this clergyman: see twenty-nine of his letters in his " Private Correspondence," written in 1788—1792, addressed to Mrs. King.]

ters; four of the latter died young, the other children survived him.[*]

At the close of this year he became a candidate for the Professorship of Botany, now vacant by the death of Mr. Bradley. Though he had now given several courses of Lectures, had restored the study of Botany in the University, and was the only person who could be called a master of the science, yet he was not the only candidate.

The following is part of the correspondence which occurred on this subject, between Mr. Martyn, and

---

[*] Professor John Martyn's family, by his first wife, Eulalia, was as follows,—

1. EULALIA, born . . . . died an infant.

2. THOMAS, born September 23, 1735, died June 4, 1825, (see the next Memoir in this volume.)

3. JOHN, a twin, born March 17, 1736-7; he married, and left one son, who died under age; he himself died at Annopalis, in Maryland.

4. EULALIA, a twin, born March 17, 1736-7, died February 14, 1737-8.

5. ELIZABETH, born December 23, 1738; she married the Rev. Daniel Longmire; by whom she had two children, Eulalia Maria, (who married her cousin, the Rev. John King Martyn, and died June 19, 1807,) and John (married and now living): she died January 4, 1825.

6. GEORGE NATHANIEL, born August 21, 1740, died unmarried, in the East Indies.

7. KATHARINE-EULALIA, born March 31, 1743, died July 11, 1747-8.

8. MARY FRANCES, born March 30, 1745, died August 5, 1746.

his friend, the Rev. Richard Arnald, Fellow of Emmanuel College, Cambridge :—

From the Rev. Mr. Arnald*, to Mr. John Martyn.

Emmanuel Coll., Nov. 30, 1732,

Dear Sir,

When first I came down to College, I was in hopes we might have carried the Botany Professorship for you without any opposition; and indeed imagined the University, out of regard to your known merit, would have made you a compliment of it; but it was not long before I heard that Mr. Goddard, of St. John's College, was determined to appear for it, at the recommendation of Dr. Middleton; who says, he talked to you, and you seemed very indifferent about it. There is now a second candidate started up, Mr. Parn, of Trinity College, and strong interest is making for him.

I find these two gentlemen are resolved to stand it out, and that many whom I have talked with are of opinion that the preference should be given to one of the Senate, if he appears for it. Upon the whole, we have thought it better to decline the further

* [RICHARD ARNALD was born in London, admitted at Corpus College, Cambridge, in 1714, and removed to Emmanuel, in 1718, where he was elected Fellow, in 1720. He resided in College till he was presented by his Society to the Rectory of Thurcaston, in Leicestershire. He died, 1756. His principal work was a Commentary on the Apocryphal Books, 1744, 1748, 1752.]

prosecution of this affair; as it is much controverted, purely honorary, and what I think you do not much set your heart upon, except you could receive it by way of compliment. . . . . .

I am, &c.,

RICHARD ARNALD.

From the same to the same.

Emmanuel Coll., Dec. 3, 1732.

Dear Sir,

Since my last letter Mr. Goddard has desisted, purely out of a sense of your superior merit. Mr. Parn we are not at all apprehensive about: he has not been at Cambridge for some time; and 'tis thought, when he comes, his friends will advise him to desist. But, be that as it will, we shall certainly have a great majority in your favour; and as the Society has this affair much at heart, they have taken all the proper steps to secure it. . . .

I am, &c.

RICHARD ARNALD.

From Mr. John Martyn to the Rev. Mr. Arnald.

Chelsea, Dec. 5, 1732.

Dear Sir,

I am greatly obliged to you and the rest of my friends for your warmth in my interest. I must own I was something surprised when I understood, by your former letter, that I was to meet with so much opposition. I thought the University would never have made any difficulty about giving me an empty title in a science which I had restored, after it had been totally lost among them, and had continued to teach for six years with much labour and little profit. I shall be always fond of any mark of their esteem for me, provided they can be unanimous in bestowing it; but never desire to be the cause of animosities amongst them: so that if it is probable there will be much opposition I desire to desist. I find it objected to me, that I have left the University, which you can refute, for you know that my name is still on the boards at Emmanuel, and that I told you, when you was here, that I had no thoughts of cutting it out. And besides, I wonder any should imagine I would show so little regard to the University as to neglect them, after such a public mark of their esteem and favour. How inconvenient soever it may be to me in the following of my practice here, which absence must necessarily injure, I shall nevertheless endeavour to serve the University whenever it can be done without the greatest detriment to my own private affairs. As to the other objection—that I will not take the oaths,—you may be assured it is entirely without foundation; and, if you remember, I intended to have taken them last summer, if I had not been obliged to leave Cambridge a week before the Assizes. . . .

I am, &c.,

JOHN MARTYN.

From the Rev. Mr. Arnald to Mr. John Martyn.

Emmanuel Coll., Dec. 17, 1732.

Dear Sir,

I thought by this post to have given you joy of the Botany Professorship; but we were advised to defer that affair 'till the beginning of next term, upon notice that our good friend, the master of Christ's, intended to stop your *Grace* in the Caput:—as you know the man, I need not tell you it was upon the surmise of your being a non-juror. I have already been with the Vice-Chancellor upon that subject, and I think satisfied him; but I wish, to remove all objections, you would take the oaths; or at least assure them by a line that you are ready to do it cheerfully if called upon. The Trinity people, upon hearing from Mr. Parn, have dropt him; so that now you are the only candidate, and have it all before you, if you will comply as above. . . .

I am, &c.,

RICHARD ARNALD.

From Mr. John Martyn to the Rev. Mr. Arnald.

Chelsea, Dec. 23, 1732.

Dear Sir,

I am sorry you have had so much trouble on my account. I cannot imagine what injury I ever did any member of the University, or what can be the occasion of so ill-natured and groundless a report as that I am unwilling to take the oaths. I am sure no word or action of mine can have given any reason for such a suspicion. The inclosed will, however, I imagine, be a sufficient answer to this objection. I must add, that I should have taken the oaths when I did, if this affair had *not* happened. . . .

I am, &c.,

JOHN MARTYN.

No man was ever a better subject than Mr. Martyn; however, to take off the aspersion which had been cast upon him, on the 12th of December he took the oaths to the King. All opposition falling at length before his superior merit, he was chosen Professor of Botany, by the unanimous voice of the University, on the 8th of February, 1733.

In the years 1732, 1733, 1734, 1735, he was engaged in "*the General Dictionary;*" in which work, however, he was concerned no further than the three first volumes: he wrote the lives of *Abdolonymus, Archimedes, Belon, Bocconi, Brunsfeld,* and some others.

In the year 1734, he published the "*Abridgement of the Philosophical Transactions, from* 1719 *to* 1732," in two volumes, 4to.—numbered Vol. VI. and Vol. VII. of the series. (See below, pp. 66, 72.) This was a continuation of the volumes which had been published by Lowthorpe and Jones. The Booksellers had engaged Mr. Eames to continue the Abridgement; and, soon after, another set put Read and Gray upon the

:same work.   Mr. Eames proceeded so slowly, that it
was probable the latter would get the start; they,
therefore, engaged Mr. Martyn, in 1732; who, while
,Eames abridged three chapters, finished all the rest!

In the year 1735 he read his last Course of Lectures in Botany at Cambridge; labouring under great
disadvantages for want of a Botanic Garden, and not
finding sufficient encouragement to warrant so long a
neglect of his practice as the Course must necessarily
occasion.

In 1737, he published his " 5th," *and last, " Decad
of rare plants.*"   [This splendid work, not having
been adequately supported, was thus brought to an
end, at the fiftieth plate.   At p. 50, there is a figure
of a " beautiful Helleborine," (the *Limodorum altum,*
or *Bletia verecunda* of the Kew Catalogue,) sent to
the Chelsea Garden by the celebrated Mr. Peter Collinson; whose letter claiming to have the Engraving
dedicated to himself, preserved among Sir Joseph
Banks's MSS., in the British Museum, shows in what
high estimation Mr. Martyn's work was held by the
first Botanists of that period.   He was strongly urged
to continue this noble publication; and, at the request of Sir Hans Sloane, (MSS. Sloane, 4052, p.
282,) he made an estimate of the number of new
subscribers which would enable him to do so; without
the loss of £50 upon each Decad, which he was then
incurring: but as he did not meet with the requisite
encouragement, the work proceeded no further.]

He now entered into correspondence with Linnæus.

[This correspondence seems to have been occasioned
by Linnæus presenting him with a copy of his " Flora
Lapponica," published in that year.   The copy is still
preserved in the Botanical Library at Cambridge,
and exhibits the autograph[a] of Linnæus at the foot
of the title page,—" *Celeb. Professori Joh. Martyno.*"
" He was," says Dr. Pulteney[b], " if not the first, at
least one of the earliest English writers, who announced the Northern Genius to the British reader:
this was done by the Professor's extract from the
' Flora Lapponica,' printed in the edition of the
' Georgics,' in 1741 : it was some years afterwards
before the system of the Swede made any progress in
England."   The extract alluded to, is to be found in
Mr. Martyn's note on the 196th line of the third
Georgic, and is that passage in which the enthusiasm
of the illustrious naturalist burst forth in that elo-
quent apostrophe—" *O felix Lappo !* " &c.[c]]

[a] [Linnæus presented him also with the first edition of his
Genera Plantarum, and his *Critica Botanica ;* inscribing them,
with his own hand, " *Clarissimo Prof. Martyno.*"]

[b] [Pulteney's Sketches of the progress of Botany in England,
Vol. ii., p. 217.]

[c] [The passage is thus introduced : " My learned friend, Dr.
Linnæus, of Upsal, who travelled in Lapland, in 1732, was pleased
to send me an excellent account of the plants of that country, under
the title of *Flora Lapponica,* printed at Amsterdam, in 1737,
in 8vo.   Speaking of a dwarf sort of birch,—the *Betula nana,*
or, *Dwarf Birch,*— which is greatly used in the Lapland œco-
nomy, he takes occasion to extol the felicity of the Laplanders.
He says, they are free from cares, contentions, and quarrels, and

He next published, "*A Treatise on the powers of Medicines, by the late learned Herman Boerhaave, Doctor of Philosophy and Physic, and Professor of Physic, Botany, and Chemistry, in the University of Leyden. By* JOHN MARTYN, F. R. S. *Professor of*

are unacquainted with envy. They lead an innocent life; continue, to a great age, free from myriads of diseases with which we are afflicted. They dwell in woods, like the birds, and neither reap nor sow.—'O felix Lappo, qui in ultimo angulo mundi sic beae lates, contentus et innocens! Tu nec times annonæ charitatem, nec Martis prælia, quæ ad tuas orás pervenire nequeunt, sed florentissimas Europæ provincias et urbes, unico momento, sæpe dejiciunt, delent. Tu dormis hîc, sub tua pelle, ab omnibus curis, contentionibus, rixis liber; ignorans quid sit invidia. Tu nulla nôsti, nisi tonantis Jovis, fulmina. Tu ducis innocentissimos tuos annos, ultra centenarium numerum, cum facili senectute et summa sanitate. Te latent myriades morborum nobis Europæis communes. Tu vivis in sylvis, avis instar, nec sementem facis, nec metis; tamen alit te Deus Optimus optimè. Tua ornamenta sunt tremula arborum folia, graminosîque laci. Tuus potus aqua chrystallinæ pelluciditatis, quæ nec cerebrum insania adficit, nec strumas in Alpibus tuis producit. Cibus tuus est, vel verno tempore piscis recens, vel æstivo serum lactis, vel autumnali tetrao, vel hyemali caro recens rangiferina, absque sale et pane, singula vice unico constans ferculo; edis dum securus e lecto surgis, dumque eum petis, nec nôsti venena nostra quæ latent sub dulci melle. Te non obruit scorbutus, nec febris intermittens, nec obesitas, nec pedagra; fibroso gaudes corpore et alacri, animoque libero. O sancta innocentia! estne hic tuus thronus, inter Faunos, in summo septentrione, inque vilissima habita terra? numne sic præfers stragula hæc betulina, mollibus serico tectis plumis?—Sic etiam credidêre veteres, nec malè.'"— See the *Flora Lapponica*, p. 269.]

Botany in the University of Cambridge. London, 1740."
8vo.

Mr. Martyn's fine edition of Virgil's Georgics,
appeared the ensuing year, [with the following title :—
" *Pub. Virgilii Maronis Georgicorum Libri Quatuor.
The Georgicks of Virgil, with an English Translation
and Notes:* By JOHN MARTYN, F.R.S., *Professor of
Botany in the University of Cambridge.* London,
1741." 4to. pp. xxii. and 404.] It is dedicated to
the famous Dr. Mead; who acknowledged this attention in the following letter :—

From Dr. Mead to Mr. John Martyn.

Ormond-Street, March 23, 1740-1.
Sir,

Ever since I received the favour of your
Virgil's Georgics, with your Letter, and the Copy of
a Dedication, I have at leasure minutes entertained
myself with reading the divine Poet, and your learned
Commentary. I could not have thought that after
so many great Critics and Commentators, so much
more could have been done to illustrate that noble
work. I dare say the learned world will own that
you have outdone them all. You have done great
honour to my two MSS., and I am glad they have
been more useful to you than I imagined they would,
being of no great age; but your good judgment and
sagacity has shown itself in finding out always the
best readings.

I am very sensible of the honour you do me, in

prefixing my name to so valuable a performance, when you might have chosen Patrons in all respects more worthy. It is always agreeable Laudari a viro laudato. But it gives me a particular pleasure on this occasion, that this will, I hope, be a foundation of an acquaintance and friendship between us, and that I may hereby have it in my power to show with how much respect,

I am, &c.,

RICHARD MEAD.

[Besides several maps, &c., this volume contains engravings* of the following plants. 1. *Citrus Medica,* 2. *Elæagnus angustifolia,* 3. *Olea Europœa,* 4. *Cerinthe major,* 5. *Lilium Martagon,* 6. *Aster Amellus ;—* (the modern names being here substituted for Mr. Martyn's.)] He was fond (perhaps to a degree of enthusiasm) of the ancients, particularly of their poets : but this poem was ever his favourite; and the general bent of his studies led him to understand many parts of it better than most men in Europe. The Astronomical part, with which he modestly professed to be least acquainted, " had the good fortune," as he expresses it, " to be perused by the greatest Astronomer of this, or perhaps of any age; the enjoyment of whose acquaintance and friendship, I shall always esteem as one of the happiest circumstances of my life. " This was no less a person than

* [The original drawings are still preserved]

the celebrated Halley, whose learning he always
venerated. Notwithstanding his own knowledge on
this subject appeared, to himself, nothing, because he
had been accustomed to consider it along with that
of his learned friend; yet he paid a just veneration
to the study, and was by no means unacquainted
with the writings of the immortal Newton. At first,
he intended that his notes should have been in Latin,
to render them of more extensive use; but, though
he had made a considerable progress, (as the MS.
which still remains may serve to testify,) yet, at length,
the natural bias for his native country, (which was
ever strong in him) prevailed, and he was content to
sacrifice a more extensive reputation to that laudable passion. He has, however, since been rewarded
for this sacrifice; his edition of this famous work
being well known and much respected in most parts
of Europe.* [We shall recur to this subject, when
we come to notice his edition of *Virgil's Bucolics*,
published in 1749.]

Some further communications to the Royal Society,
appear in the volume for 1741; viz.

1. *An Account of a new purging Spring, discovered
at Dulwich, in Surrey: by Mr.* JOHN MARTYN, F.R.S.,
Prof. Botan. Cantab." (Phil. Trans. for 1741, No.
461, Vol. XLI., Part II., pp. 835—838.) [This spring
was discovered in 1740, on digging a well, sixty feet

* [A German translation of it was published at Hamburg, 8vo.
in 1759.]

deep, at the Inn, called "the Green Man." Mr. Martyn gives a particular account of the strata cut through, and of the quality of the water.]

2. *" A Letter, from Mr.* JOHN MARTYN, F. R. S. *Prof. Botan. Cantab., to John Machin, Esq., Sec. R. S., and Prof. Astron. Gresham ; concerning an Aurora Australis, seen March* 18, 1738-9, *at Chelsea, near London."* (Phil. Trans. for 1741, No. 461, Vol. XLI., Part II., pp. 840-842.) [An extraordinary redness in the air first attracted Mr. Martyn's attention, which was taken by country people for a great fire towards London. " Fiery red rays or bands" then shot down from the zenith to the south-west, " till the whole southern atmosphere was tinged of a red brightness."]

The latter end of this year he set about the translation and abridgment of *" the Philosophical History and Memoirs of the Royal Academy of Sciences at Paris : or an Abridgement of all the papers relating to Natural Philosophy, which have been published by the Members of that illustrious Society. Illustrated with copper-plates. The whole translated and abridged by* JOHN MARTYN, F.R.S., *Professor of Botany in the University of Cambridge ; and* EPHRAIM CHAMBERS, F. R. S., *Author of the Universal Dictionary of Arts and Sciences. London,* 1742." 8vo. 5 volumes.—[He was much disappointed with regard to the assistance he had hoped to derive from his colleague in this work ; as appears by the following letter.]

From Mr. John Martyn to Mr. Knapton.

Chelsey, May 6, 1742.

Sir,

When I saw you last, I told you that the declension of the sale of our "*Abridgment*," was entirely owing to the papers translated by Mr. Chambers. To prove my assertion, I have sent you the first paper that came in my way; which I assure you is very far from being one of the worst; and desire you would give yourself the trouble to look over it. The author writes with great prolixity; and Mr. Chambers is so far from having abridged him, that he has paraphrased him; sometimes using two or three words, where the Author contents himself with one. But what is worse, the English style (if it may be called English) is very low and poor, and full of blunders. I shall only point out to you some of the errors of a few pages, which I wish you would read over carefully, and try if you can even make sense of them. P. 160. "*Remarkables are discoverable ;*" this sounds very ill to the ear, and I question whether "*remarkables*" is English: in the French, it is "*ce qu'il y a de singulier.*" "I have extended my observations to *see-nettles*," instead of "*sea-nettles.*"—P. 161. "*folding,* or *two-leaved*," might be expressed by one proper word, *bivalve.*—P. 165. "*Lavenion,*" instead of the French word "*Lavignon.*"—Mr. Chambers was ignorant of the English names of most of the shell-fishes. Thus he translated *"œil de bouc"* the *"goat's-eye,"* instead of *limpet,* which is a well-known name; and these *" Lavignons,"* are called *Purrs,* on our coast.—I shall trouble you with no more; the paper goes on in the same manner, or rather worse. I will only desire you to turn to page 182, where you will find my hand again to some French words, spelt in a surprising manner.—Mr. Chambers never makes use of any stops; which occasions a great deal of trouble both to the printer and me: most of his papers are so ill done, that it would be as little trouble to translate them from the original, as to reduce his to common sense and tolerable English. . . .

I am, &c.,

JOHN MARTYN.

His next work was, *" A Treatise of the acute diseases of Infants: to which are added, Medical Observations on several grievous diseases. Written originally in Latin, by the late learned* WALTER HARRIS, *M. D., Fellow of the College of Physicians at London, and Professor of Chirurgery, in the same College. Translated into English, by* JOHN MARTYN, *F.R.S., Professor of Botany, in the University of Cambridge. London, 1742."* 8vo., pp. xiv. and 228.

On the 8th of April, 1743, he lost his respected father, who had nearly attained the age of eighty-one.

Another very curious paper now appeared in the Transactions of the Royal Society:—*" An Account of*

a new species of Fungus: by JOHN MARTYN, F. R. S.,
Prof. Bot. Cantab." With a plate. (Phil. Trans. for
1745, No. 475, Vol. XLIII., pp. 263, 264.) [This ex-
traordinay Fungus, was found on an elm log, in a damp
cellar, in the Hay-Market, London. It is the *Boletus
rangiferinus,* or *Rein Deer Boletus,* of Withering. It
is figured, not only in the Phil. Trans., but also in a
detached folio plate by Ehret,—in Blackstone's Spec.
Botanicum,—in Bolton's Hist. of Fungi about Halifax,
t. 138,—and in Sowerby's Fungi, t. 266, under the
name of *Boletus squamosus :* the latter, however, (*if* it
be the same Fungus,) is a very different variety from
Mr. Martyn's plant.]

In 1747, he published, in two vols., 4to., the
" *Abridgment of the Philosophical Transactions, from
1732 to 1744,*" being Vol. VIII. and Vol. IX. of the
series. He had been engaged in this work from the close
of the year 1733. (See above, p. 56, and below, p. 72.)

In 1749, he printed his translation and notes upon
the Bucolics of Virgil ;—[" *Publii Virgilii Maronis
Bucolicorum Eclogæ decem. The Bucolicks of Virgil,
with an English Translation and Notes.* By JOHN
MARTYN, F. R. S., *Professor of Botany in the University
of Cambridge. London,* 1749." 4to. pp. lxv. and 280.
Besides two maps, there is a plate of the following
plants in a group :—1. *Lilium album ;* 2. *Cheiranthus
fruticulosus ;* 3. *Papaver Rhœas ;* 4. *Narcissus poeticus ;*
5. *Anethum graveolens ;* 6. *Daphne gnidium ;* 7. *Lilium
Martagon ;* 8. *Calendula officinalis ;*—(the modern
names being here substituted for Mr. Martyn's.)

There is also a well engraved bust of Virgil, from an antique gem.] He immediately went on with the work, designing to complete the edition of this favourite poet ; but growing infirmities too soon stopped his hand, and prevented him from accomplishing more than a few Dissertations and Critical Notes. [These were left behind him in MS., and appeared, as a posthumous publication,* (to which this Memoir was originally prefixed,) with the following title :—" *Dissertations and critical remarks upon the Æneids of Virgil. Containing, among other interesting particulars, a full vindication of the Poet from the charge of an anachronism, with regard to the foundation of Carthage. By the late* JOHN MARTYN, F. R. S., *Editor of Virgil's Georgicks and Bucolicks. London,* 1770." 12mo. pp. 227. This work, being confessedly a fragment, has not been reprinted ; although it might be advantageously added as a Supplement to the two other Volumes. The " *Georgics*" and " *Bucolics*" have

* [This publication was dedicated by Professor Thomas Martyn to Daniel Crespin, Esq. The Dedication contains the following passage, with reference to the merits of the work. " You, Sir, who have not merely trod on classic ground, but for many years have breathed classic air, will gladly accept these attempts to vindicate and illustrate one of the finest Authors of antiquity. Had diseases and infirmities spared our Friend a few years longer, these Essays would have risen into a complete edition of Virgil. However, imperfect as they are, you will peruse them with pleasure ; and the public, it may be presumed, will not disdain to receive illustrations of the Æneids, from the same pen which first made the Georgics of Virgil well understood."]

passed through many editions; of the former, the fifth impression appeared in 1827 ; of the latter, the fourth, in 1820. The reprints, since Mr. Martyn's death, being conducted exclusively by the booksellers, are grossly inaccurate. It is surprising, also, that no scientific editor has hitherto been employed to give to this valuable Botanical Commentary on Virgil a more acceptable form, by annexing the Linnæan names of the plants to the inconvenient and clumsy phraseology which obtained before the Linnæan language had been introduced into England. But, notwithstanding such defects, these volumes are invaluable. " To the classical reader in general," observes Dr. Pulteney,* " they afford ample satisfaction ; but to those who join to such elegant enjoyment a knowledge of the learned Editor's favourite science, these volumes must afford a gratification which they will in vain seek for elsewhere. His great knowledge, both of ancient and modern science, enabled him to appropriate the modern appellations, with a degree of judgment that has been highly approved of by those who know the difficulty of the undertaking, under that almost total want of specific distinction which occurs in the writings of the ancients." Nearly 130 plants occur in the Bucolics and Georgics : to determine the modern names of which, is the object of Professor Martyn's elaborate criticisms ; in fact, these volumes present us

* [Dr. Pulteney's Sketches of the progress of Botany in England, Vol. II., p. 217.]

with a complete Flora Virgiliana, as far as the Bucolics and Georgics are concerned.]

This year, (1749,) he lost his bosom friend, by a cancer in her breast, occasioned by a violent blow given her as she was walking in London :—having lived with her almost twenty years, with the most unremitted affection, it is no wonder if he suffered severely by the loss.

Several communications were made by him, the next year, to the Royal Society, viz :—

1. *" An Account of an Aurora Australis, seen Jan. 23, 1749-50, at Chelsea: by* JOHN MARTYN, F. R. S., *Prof. Bot. Cantab."* (Phil. Trans. for 1750, No. 494, Vol. XLVI., pp. 319—321.)

2. *A Letter from* Mr. JOHN MARTYN, M. D.*, *Prof. Botan. Cantab., and F. R. S., concerning an Aurora Borealis, seen Feb.* 16, 1749-50." (Phil. Trans. for 1750, No. 494, Vol. XLVI., p. 345.)

3. *" An Account of an Earthquake felt at London, Feb.* 8, 1749-50, *in a Letter from* JOHN MARTYN, M. D., *to the President."* (Phil. Trans. for 1750, No. 497, Vol. XLVI., pp. 609, 610.—[This earthquake was felt in London, Chelsea, and about two miles more to the West, at noon. Mr. Martyn, and several other writers in the Transactions, concur in representing the effect as similar to that arising from some heavy

* [It does not appear that Professor John Martyn ever took the degree of M. D. It was assigned to him by mistake, in this and the two following papers, in the Phil. Trans.]

weight thrown suddenly down, shaking the floors, &c. &c.]

4. " *An Account of an Earthquake felt at London, March 8, 1749-50: in a Letter from* JOHN MARTYN, M. D., F. R. S., *to the President.*" (Phil. Trans. for 1750, No. 497, Vol. XLVI., pp. 630-631.) — [This earthquake was much stronger than that which occurred, Feb. 8;—" those who were out of doors felt the ground shake under them very sensibly ; whereas, in the former, few were sensible of the shock, except those who were in houses."]

In July, 1750, he married MARY ANNE, daughter of CLAUDE FONNEREAU, of London, merchant, who bore him one son, Claudius,* and survived him.

Soon after this, in the spring of the year 1752, he retired from practice and the world, to a delightful farm, situated on a hill, remarkable for its rich and extensive prospects, in the parish of Streatham, in Surrey. Here, for the remainder of his life, he enjoyed the sweets of rural retirement; as far as frequent attacks of the gout in his head and stomach would permit; and rejoiced to close his days with the practice of that art which his favourite Poet had sung so well, and of which the Roman Orator says—" si non est sapientia, est proxima tamen sapientiæ."

We have now to notice his *last* communication to the Royal Society.—" *A remark concerning the Sex of*

---

* [The Rev. CLAUDIUS MARTYN, was Rector of Luggershall, Bucks. He died 1828.]

Holly: By Mr. John Martyn, F. R. S., *Professor of
Botany in the University of Cambridge.* (Phil. Trans.
for 1754, Vol. XLVIII., Part II., pp. 613—616.)—This
was a new discovery.—[The *Ilex aquifolium,* or *Common Holly,* had hitherto been supposed to bear only
hermaphrodite flowers, by Ray and other Botanists·
But Mr. Martyn, by close observation of some plants
in his garden at Streatham, discovered that it was
diœcious; and, consequently, maintained that it ought
to be removed from the Linnæan class, *Tetrandria
tetragynia,* to that of *Diœcia tetrandria.* The Royal Society, however, referred the subject to Dr. (afterwards
Sir William) Watson, who confirmed Mr. Martyn's discovery; but found that the Holly was very variable in
its habit; different plants being hermaphrodite, monœci-
ous, and diœcious : he, therefore, decided that its proper class is, *Polygamia triœcia.* Notwithstanding these
remarks, modern Botanists have judiciously kept the
Holly in its old class, *Tetrandria tetragynia ;* both because (even when the plant *is* diœcious) there is no
real difference between the accessary parts of the male
and female flowers, and because the diœcious character does not appear in other species of this genus.
The class of *Ilex aquifolium* is, therefore, determined
upon the same principles as those of *Valeriana dioica,
Lychnis dioica,* and many other anomalous species.
It is remarkable, however, that the variable habit of
the *Holly* is not noticed in Dr. Smith's English Flora;
the fact being, undoubtedly, a curious one.]

　　He now completed his engagement with the Booksellers for the "*Abridgment of the Philosophical Trans-actions, from* 1743 *to* 1750," by the publication, in 1756, of Vol. X., of the series. (See above, pp. 56, 66.) This volume, of the Abridgment, corresponds with Vol. XLVI., of the Original; when the Royal Society, on the death of their Secretary, Cromwell Mortimer, (in 1752,) took the publication of their papers into their own hands.

January 30th, 1762, Mr. Martyn resigned his Pro-fessorship of Botany. Some time after, in gratitude for the favours done him, in chosing him, and after-wards his son, into this post, he presented to the University, which he had ever loved with an affection truly filial, his Library* of Botanical books, amounting to above 200 volumes; his Hortus Siccus of foreign plants, containing upwards of 2,600 specimens; nearly 250 curious drawings of Fungi; his collection of Seeds and Seed-Vessels, and his Materia Medica. The fol-lowing letter accompanied this gift:—

To the Right Worshipful the Vice-Chancellor of Cambridge.

Streatham, 1765.

Sir,

Sensible of the great honour done me by the University, many years ago, in conferring upon me the title of their Professor of Botany; and of the continu-

* [A portion of Mr. Martyn's Library was sold by auction, after his death, in 1768, by Lockyer Davis and Charles Reymers. See Gough's List of Sale Catalogues, in Nichols's Lit. Anec., Vol. III., p. 637.]

ance of their favour, in bestowing the same on my son, upon my resignation ; I am desirous of making some acknowledgment.   I therefore request the Chancellor, Masters, and Scholars of the University of Cambridge, to accept of my printed books relating to Botany, and of my *Hortus Siccus,* containing a considerable number of dried specimens of plants, collected from most parts of the known world.   The whole is contained in eight wainscot cases, which I desire may be deposited in some convenient place * for the use and under the care of the present Professor of Botany, and of his successors for the time being.   I hope, Sir, the University will not disdain to receive this small donation from their most obliged, &c.

                                      JOHN MARTYN.

June 29, 1765.—It was decreed by the Senate of the University, that thanks should be given to Mr. Martyn, by the Vice-Chancellor, for this present : and, accordingly, Dr. Barnardiston, Master of Bene't College, being then in that office, wrote him a handsome letter upon the occasion.

* [The Botanical Library and Museum was, for more than fifty years afterwards, sadly exposed to injury, by the University neglecting to appropriate a convenient place for its preservation. Professor Thomas Martyn made many fruitless applications on this subject. At length, on the representations of Professor Henslow, the Senate passed a Grace in 1827, for allowing an annual stipend for the support and increase of the Museum.]

About a year, however, before his death, infirmities, arising from repeated shocks of his disorder, were so much increased, that he could no longer enjoy the delights of his farm; he removed, therefore, once more, to his house at Chelsea; and, by the most gradual decay, paid his last debt to nature, on the 29th of January, 1768. He was interred on the north side of the burying ground belonging to that parish, in the same grave with his wife and three of his children; [over which lies a flat stone, with a simple inscription. (See Faulkner's History of Chelsea, p. 106, 8vo., 1810.)]

Mr. Martyn was religious, without bigotry; devout, without superstition; charitable, without ostentation. His religion was neither founded on the mere prejudices of education, nor adopted without choice. He was, through life, a decided member of the Church of England. His regard for the Government was strong, and his patriotism ardent. His friendships were sincere, warm, and permanent. His benevolence was unconfined; but it was *most* conspicuous in an unremitted attendance on the maladies of *the poor*. He was always willing to make considerable sacrifices at the shrine of peace; and often congratulated himself that he had never been obliged to appear in any of the Courts of Law either as principal, or evidence. The warmth of his affection as a husband, and his truly parental love were most exemplary.

His literary accomplishments must be left, in a great

measure, to be tried by the books which he has published. His knowledge in Botany* will, from them, stand confessed; but, notwithstanding Natural History was ever his chief delight and employment, yet he by no means confined himself to that alone. He attended with sedulity to all that knowledge which is requisite to make a good Physician; and in his profession had peculiar success in treating the small-pox, and nervous disorders.

His classical abilities were considerable; his relish for the ancient poets never forsook him; and even after he had lost the power of reading, he could hear the odes of Horace with that pleasure, which is usually confined to the earlier part of life.

To his acquaintance with the ancient, he added also the modern languages: his own he studied critically; and had actually composed a Grammar of it, and had made large collections towards a new *English Dictionary*, upon the same plan with Dr. Johnson's. History, both ancient and modern, together with its appendages, Biography, Geography, Chronology, and Genealogy, he was a perfect master of; Divinity, and Ecclesiastical History, were the studies of his latter days.

---

* That his knowledge of Botany might not be looked upon as purely speculative, he introduced into practice, *Valerian, Pepper-Mint Water*, and *Black Currants*,—three medicines with which, under his own eye, he wrought effects much to the credit of his profession.

His Epistolary Correspondence with learned men
was very extensive for about twenty years. During
that time he corresponded with Dr. Blair[a]; Mr. Wil-
mer[b]; Mr. Francis Dale; Mr. Tullidelph[c]; Mr. P.
Miller; Dr. Richmond, of Burton, near Wootton-
Basset, in Wiltshire; Dr. Richardson[d], of North-
Bierley, in Yorkshire; Mr. L'Isle; Dr. Douglas; Mr.
Hawksbee; Mr. W. M. Gilkes[e]; Mr. Richard Davies,
of Queen's College, in Cambridge; Mr. Brewer, of
Bradford, in Yorkshire; Sir Hans Sloane; Dr.
Fysher, Fellow of Oriel College, in Oxford; Dr.
Thomas Dale; Mr. Halfhyde; Dr. Russell; Dr.
Massey; Mr. Houstoun[f], Mr. Arnald[g], Beaupré Bell,
junior, Esq., of Beaupré Hall, in Norfolk; Dr. Rich-
ard Rawlinson; Mr. Holmes, Fellow of Emmanuel
College; Mr. Samuel Dale; Mr. Blackstone[h]; Mr.
Peter Collinson[i]; &c.: also with Signor Micheli, of
Florence; Dr. Boerhaave, and John Frederic Grono-

[a] See above, p. 7.   [b] See above, p. 7.   [c] See above, p. 19.

[d] [There is a long account, with a portrait, of Dr. Richardson,
and a good deal of his correspondence, in Nichols's Illustr. of Lit.
Anec., of 18th century, Vol. I., p. 225-416. He was born, 1663,
and died, 1741.]

[e] [Some interesting letters addressed to Mr. Martyn by Mr.
Gilkes, from Italy, in 1744-5, are in the British Museum, among
the papers given by Professor Thomas Martyn to Sir Joseph
Banks.]

[f] See above, p. 46.   [g] See above, p. 52.

[h] [Author of the "Specimen Botanicum;" or places of native
growth of many rare British Plants, 1746.]

[i] See above, p. 57.

vius, of Leyden ; M. Jussieu, of Paris ; and the celebrated Linnæus.

His knowledge was solid, though extensive; and he
was learned without pedantry. Though his application had been severe, yet no man possessed a greater
relish for social converse, or the intercourse of friendship. [His literary and philosophical character has
been thus summed up by one who afterwards trod in
the same path, and was fully competent to appreciate
his merits. "It is not without the strictest justice,"
says the late Dr. Pulteney,* "that the term indefatigable is applied to this learned man. His avocations
from business were wholly devoted to the cause of
literature ; which he contributed to serve in various
ways. Besides the obligations which literature in
general owes to this learned professor, that which I
call more strictly *English* Botany, received considerable augmentation from his labours."]

* [Pulteney's Historical Sketches of the progress of Botany in
England. Vol. II., pp. 215—217.]

LIST OF THE PRINTED AND MANUSCRIPT WORKS OF
PROFESSOR JOHN MARTYN.

I. Published Works[a].

1. A poetical translation of a Latin Ode, by Camerarius, on the sexes of plants. 8vo. 1720.

2. Tabulæ Synopticæ Plantarum Officinalium. Folio, 1726.

3. Methodus Plantarum circa Cantabrigiam nascentium. 12mo. 1727.

4. Historia Plantarum Rariorum. Folio. 50 plates. 1728 —1737.

5. Observations in Natural History, in a journey to the Peak, in Derbyshire. Phil. Trans., Vol. XXXVI., for 1729.

6. The first of a Course of Lectures in Botany, given at Cambridge. 8vo. 1729.

7. Papers communicated by B. to the Grub-street Journal, in 1730—1737.

8. Translation of a Memoir by M. Garcin, on the Genus *Oxalis*, and *Musa*. Phil. Trans., Vol. XXXVI., for 1730.

9. Translation of Tournefort's History of Plants growing about Paris, with additions; accommodated to Plants growing in Great Britain. 2 vols. 8vo. 1732.

10. Abridgment of the Philosophical Transactions, from 1720 to 1732. 2 vols. 4to. (viz. Vol. VI. and Vol. VII.) 1734.

[a] [The titles are here given in a compendious form; as a more full account of each work may be seen in the preceding Memoir, under the date of each publication.]

11. Some Articles furnished for the first three Volumes of "The General Dictionary"; viz. the Lives of Abdolonymus, Archimedes, Belon, Bocconi, Brunsfeld, &c., 1732—5.

12. Translation of Boerhaave on the powers of Medicines. 8vo. 1740.

13. The Georgics of Virgil, with an English Translation, and Notes. 4to. 1741.

14. Account of a new purging Spring at Dulwich. Phil. Trans., Vol. XLI., Part II., for 1741.

15. Account of an Aurora Australis, seen 1739, at Chelsea. Phil. Trans., Vol. XLI., Part II., for 1741.

16. Abridgment of the Memoirs of the Royal Academy of Sciences, at Paris, (published in conjunction with Mr. Chambers). 5 vols. 8vo. 1742.

17. Translation of Harris on the Diseases of Infants. 8vo. 1742.

18. Account of a new Fungus, *Boletus rangiferinus.* Phil. Trans., Vol. XLIII., for 1745.

19. Abridgment of the Philosophical Transactions from 1732 to 1743. 2 vols. 4to. (viz. Vol. VIII., Vol. IX.) 1747.

20. The Bucolics of Virgil, with an English Translation, and Notes. 4to. 1749.

21. Account of an Aurora Australis, seen at Chelsea, 1750. Phil. Trans., Vol. XLVI., for 1750.

22. Account of an Aurora Borealis, seen at Chelsea, 1750. Phil. Trans., Vol. XLVI., for 1750.

23. Account of an Earthquake, felt in London, Feb. 1750. Phil. Trans., Vol. XLVI., for 1750.

24. Account of an Earthquake, felt in London, March, 1750. Phil. Trans. Vol. XLVI., for 1750.

25. On the diœcious character of Holly. Phil. Trans., Vol. XLVIII., Part II., for 1754.

26. Abridgment of the Philosophical Transactions, from 1743 to 1750; one vol. 4to. (viz. Vol. X.) 1756.

27. Nineteen Dissertations, and some Critical Remarks on the Æneid of Virgil, 12mo., (Posthumous) 1770.

II. Unpublished Manuscripts.

1. A new Methodical Distribution of Plants; containing their Classical and Generical Characters.

2. Miscellaneous Descriptions of Plants.

3. A Course of Lectures on Botany.

4. Two Common-Place Books, in large folio; containing what is to be collected from the Ancients, concerning Plants. (*The Greek are in the 1st, and the Latin Writers in the 2d Volume.*)

5. Index Plantarum circa Londinum provenientium.

6. A Methodical Distribution of Fossils.

7. Lectures on the Materia Medica.

8. Practice of Physic; in one volume folio, and one 4to.

9. A Grammar of the English Language.

10. Collections for a Dictionary of the English Language, (*on the plan afterwards followed by Dr. Johnson.*)

11. Julius Cæsar's Expedition into Britain, extracted from his Commentaries.—(*This MS. contains also the entire history of the Romans in Britain, extracted from the ancients.*)

12. The first book of Herodotus's History, translated into English.

13. Collections for an Universal Biographical Dictionary, in three volumes, folio.

14. An Introduction to Geography.

15. Genealogical Tables of the Kings and Nobility of England, &c., in upwards of 200 folio pages.

16. Chronological Tables: compiled chiefly with a view to settle the Sacred Chronology, and to connect it with the Profane.

17. Collections for a History of the Royal Society.

18. Commentaries in Latin, upon the Georgics of Virgil: two volumes, folio.

19. Commentaries on the Æneid of Virgil.

The Autograph of John Martyn, 1739, Æt. 40.

Memoir

OF

THOMAS MARTYN, B.D., F.R.S., F.L.S.,

BY

GEORGE CORNELIUS GORHAM, B.D.

MEMOIR

OF

THOMAS MARTYN, B.D., F.R.S., F.L.S.

&c. &c.

THOMAS MARTYN, the eldest child of JOHN
MARTYN, (the subject of the preceding Memoir,) by
EULALIA, his first wife, was born in Church-Lane,
Chelsea, on the 23d of September, 1735, (O. S.,) in
the house which had been the birth-place of his mo-
ther, and the residence of his grand-father, the Rev.
Dr. King, Rector of that Parish.

At a very early age, (according to the best of his
own recollections, before he was six years old,) he was
sent to school under the Rev. Mr. Rothery; who had
an establishment in Paradise-Row, Chelsea. Mr.
Rothery had been educated at Westminster school,
whence he removed to St. John's College, Cambridge:
he was a classical scholar; but his attainments never
produced him any other preferment than the Lectureship of Chelsea, and Ebury Chapel, in the Five-Fields.
Adverting to this early period of his life, at the advanced age of eighty-six, Mr. Martyn observes,—"for
about ten years, I walked six times a day between

Church-Lane and Paradise-Row; I knew and was known to almost every body in Chelsea, which has of late years rendered it a melancholy walk to me,— knowing and being known of nobody!" At this seminary he continued as a day-scholar, till he went to the University; " reaping as much instruction from his learned father at home, as from his master at school." He lived, during this period, with great domestic comfort; and the foundation of his future eminence was, undoubtedly, laid in a solid education, imparted by a competent tutor under a father's vigilant eye. While the sterner intellectual faculties were thus early developed, the more gentle dispositions and softer qualities of his naturally amiable mind were not left uncultivated : for his youthful character was formed by the care of a mother, whom he describes as being " singularly sweet, even-tempered, and indulgent." He had the misfortune to lose this watchful guardian of his tender years, when he was scarely thirteen; she died* on the 13th of February, 1748-9.

The following reminiscences of the period and scene of Mr. Martyn's earliest days, which carry us back almost a century, will not be perused without that melancholy interest which attaches to a momentary glance at objects which have long since passed away.

" Winchester House[b], and the Manor House, which was contiguous, and in which I well knew Sir Hans

* See above, p. 69.

Winchester House was pulled down in 1813.

Sloane lived near fifty,"—it may now be said ninety
" years ago, are on the ancient site of Henry VIII's
Palace. No part of the *original* building now remains. After Sir Hans's death, the Manor House
was pulled down, in 1740, and a row of houses built
in front of it nearer the river. . . . When my grand-
father, who was Rector of Chelsea, lived in Church-
Lane, *he had a clear view to London;* nothing being
then interposed but the Royal Palace, and a house or
two of the nobility, on the brink of the river; of
these, one was [Shrewsbury House, long erroneously
supposed[a] to be] Lord Chancellor More's."[b]

" I beg leave to consider Sir Hans Sloane as one of
my patrons. The condescension of the venerable and
amiable old gentleman to me, *when a school-boy,* will
never be forgotten by me. His figure is, even now,
presented to my eye, in the most lively manner; as he
was sitting fixed by age and infirmity in his arm-chair.

---

[a] [The words in brackets are intended to correct a popular error,
(into which Mr. Martyn himself fell) that Shrewsbury House was
originally Sir Thomas More's. This mistake was either originated
or confirmed by Hearne, in 1716, in his note to a passage of Roper's
Life of Sir T. Moore, (p. xxx.,) but was corrected by Dr. King, in
a letter to Hearne, existing in MSS. Sloane, No. 4455, p. 20—21.
Sir T. More's house actually stood at the north end of what is
now Beaufort Row; it was called Beaufort-house when it was purchased by Sir Hans Sloane, in 1736; it was pulled down in 1740.
Erasmus, writing to Bishop Smith, of Vienna, says, of Sir T. More's
house, " extruxit ad flumen Thamisin, haud procul ab urbe Lon-
dino, Prætorium, nec sordidum, nec ad invidiam usque magnificum,
commodum tamen."]

[b] Letter from Professor Martyn to Miss Hawkins, 12 Aug., 1790.

I usually carried a present from my father of some book that he had published, and the old gentleman in return always presented me with a broad piece of gold, treated me with some chocolate, and sent me with his librarian to see some of his curiosities. *It appears now* [1821] like looking into other times!"[*]

When Mr. Martyn was in his seventeenth year, he was admitted a pensioner of Emmanuel College, Cambridge, (June 24, 1752,) under the care of the Rev. Henry Hubbard, B. D., the tutor. This very learned man proved a steady friend to him, while he was pursuing his studies at the University; and used his influence successfully to obtain promotion for his deserving pupil. He was matriculated, December 18th, 1752; obtained a scholarship on Dr. Whichcot's foundation, in January, 1753; another on the foundation of the College, February 12th, 1755; and one of Dr. Thorpe's exhibitions, on the 8th of July following.

On January 23d, 1756, he proceeded to the degree of B.A.; on which occasion his tutor, Mr. Hubbard, congratulated his father, by letter, on the creditable manner in which his son had passed his examination for University honours. In truth, he had spent his time very diligently at College. He did not, indeed, obtain the most elevated distinctions, which that learned body has to confer on its candidates for de-

[*] Letter from Professor Martyn to Mr. Faulkner, of Chelsea, May 19, 1821.

grees, being only the fifth in the rank of *Senior Op-*
timès; for he had no taste for the abstract sciences, and
for those higher branches of the pure Mathamatics, a
considerable acquaintance with which, is essential to a
place in the first class of honours at Cambridge.   By
no means despising analytical studies, he felt himself
inclined to pursuits more immediately connected with
facts and daily observation.   When he took his degree, he had, however, made a tolerable proficiency
in Physics ; and it had been his chief aim, during the
usual three years' period of academical study, to
acquire as much *general knowledge* as possible.

Being now above twenty years of age, he applied
himself principally to preparation for Holy Orders.
At the same time, by way of relaxation, he resumed
the classical studies in which he had been so well
grounded before his entrance at the University ; and
read with ardour the Greek Historians,—Herodotus,
Thucydides, and Xenophon : in these he was encouraged and assisted by Mr. Hurd, then Fellow of
Emmanuel, and afterwards Bishop of Worcester.
That he might pursue his studies more at leisure, his
father indulged him with residing much in College, till
he was of an age to take Deacon's Orders.

Much as Mr. Martyn was esteemed by the Society
of his own College, and desirous as his friend and
tutor, Mr. Hubbard, was to promote his interests, circumstances were such as to shut up his hopes of
promotion at Emmanuel.   By the Statutes of that
College, he was precluded from a Fellowship.   Two

persons from the same county are not eligible at the same time; an absurd restriction, which still obtains in some other Colleges. Mr. Richardson, son of the Master, a native of Middlesex, having been recently elected, there was no room for Mr. Martyn : here, however, his kind tutor stepped in for his benefit, and recommended him to Dr. Parris, the Master of Sidney-Sussex College; where no such restriction existed, and where a remarkable opening presented itself for deserving candidates from other Societies. The circumstances were these :—the Collegiate edifices of Sidney having, for some time past, been considerably dilapidated, and the foundress having neglected to make provision for repairs, it became necessary to recur to the temporary sequestration of Fellowships for that purpose. The buildings having been thus, at length, restored, a new Society, in a manner, was to be established; for only three Fellows, of the old stock, remained; and there were but few young men properly qualified among the existing Students. Thus it became necessary to seek for aliens of the most distinguished merit. Accordingly, on the 27th of April, 1758, Mr. William Elliston was invited from St. John's College, Mr. John Hey from Catharine Hall, Mr. Hughes from Jesus, and Mr. Martyn from Emmanuel; at the same time with these aliens, two denizens were also elected, viz., Mr. Harness, and Mr. George Wollaston. With what discretion the Society acted in this election, may be gathered from the fact, that *three*, out of the *six* new Fellows, were thought worthy of still higher distinction in the University. Mr. Elliston became Master of Sidney, on the death of Dr. Parris; and, among other essential services to the College, created a fund sufficient to obviate the necessity of again sequestrating Fellowships for repairs; except under extraordinary circumstances.—Mr. John Hey* was, for many years, a sedulous tutor of the College, and the first Norrisian Professor of Divinity, from 1780 to 1795.—Mr. Martyn, as will be seen hereafter, became the Professor of Botany. The Fellowship to which he was appointed, was one of those on the foundation of the Lady Frances, Countess of Sidney. The invitation was announced to him by his tutor, in the following kind of manner:—

From the Rev. Mr. Hubbard to Mr. Thomas Martyn.

Emman. Coll., Cambr., Jan. 16, 1758.
" Sir,

"An affair for your advantage has been under consideration above these two months, and seems now to be drawing towards a conclusion. I never mentioned a word of it to you, because I was unwilling to raise any hopes till I was pretty sure of gratifying them. I can now venture to ask, whether you would accept of a Fellowship at *Sidney*. The Fellowships require a good deal of residence, which possibly may not be disagreeable to you, who, if I am not mistaken, do not dislike an University life. Some knowledge of Hebrew is necessary: the rest of the

* Dr. John Hey died in March, 1815.

examination is usually in Philosophy, Aristotle's Rhetoric, some part of the first six books in Homer, and Virgil's Georgics. It will be with the unanimous consent of the Master and Fellows; so that you will be under no apprehension of any competition, dispute, or quarrel. I have nothing further to say, than to desire your answer to this proposal, as soon as you have well considered it; and to assure you that, had your county been open here, you would have been the last person that should have been recommended to another College by

     " Your affectionate tutor and friend,

                H. HUBBARD."

     " P. S.—I ventured to say, that you are intended for Orders."

To this letter, so full of friendly feeling, Mr. Martyn made the following reply:—

From Mr. Thomas Martyn to the Rev. H. Hubbard.

            Chelsea, January 19th, 1758.

" Rev. Sir,

     " The many signal favours I have received from you at Emmanuel, will never be remembered by me without the warmest gratitude; particularly this last, so unexpected and unlooked for. Indeed, I must confess, it will be with great reluctance that I shall part with Emmanuel,—the plentiful source of all these favours, where I have received the first rudiments of science, and where there are so many persons whom

I respect or esteem. But your very kind and obliging offer, Sir, is too good not to be thankfully and joyfully accepted;—at least by one who, if he is not mistaken in himself, would of all things delight in a College life, and when my father's advice concurs entirely with my inclinations.

I propose to be at Cambridge in a fortnight at the furthest; but would come sooner, if you should think it necessary or expedient. My time here has chiefly been employed in preparing myself for Orders on Trinity Sunday; so that you may safely inform the Members of Sidney of my intention. I shall take care to comply with your commands, and to be guided by your advice in this affair. I shall likewise do my best endeavours so to prepare and behave myself, as neither to make you ashamed that you recommended, nor the Society of Sidney repent that they received,

Your dutiful pupil,

And very humble servant,

THO. MARTYN.

This event determined the course of his future life; not only by fixing him in College, for nearly sixteen years, but by bringing him acquainted with the Lady with whom he spent the remainder of his life in domestic happiness.

Soon after his election to his Fellowship, viz., May 21, 1758, he was admitted to Deacon's Orders, by the Right Rev. John Thomas, Bishop of Lincoln, in Conduit-street Chapel, St. George's, Hanover Square.

At the beginning of the year 1759, he was an unsuccessful candidate for the Lectureship at Chelsea, his native place. Though Mr. Martyn was unsuccessful, he was much beloved; by the poorer classes especially;—the elderly people crowded around him in the Church-Porch, (as he passed through them to supply Mr. Rothery's place during his illness, in 1758—9,)—adverted to their respect for the memory of his grandfather, Dr. King,—and exclaimed, " God bless you! God bless you!" " It is singular," Mr. Martyn observes, in a letter written late in life, " that *two* members of a family, so respected as mine was at Chelsea, should not be able to carry a popular election;—my uncle, Mr. King, had lost his election on a former similar occasion." However, he was accustomed to regard this failure as a providential disappointment; and he was more than reconciled to it.

July 3d, of this year, he proceeded to the degree of M.A.; December 3d, he was ordained Priest, by the above-named Bishop of Lincoln, at Buckden Palace.

Between the time of his election to a Fellowship, in the spring of 1758, and that of taking his M.A. degree, in the summer of 1759, he resided but little in College; no profit arising from his Fellowship during that period, by the limitations of the Statutes. During that interval, he lived chiefly with his father, at Streatham. When he became M.A., he was ob-

liged, by the same Statutes, to reside in College nine
months in the year;—accordingly, he returned to the
University.

A new prospect now opened before him in College,
occasioned by some remarkably rapid changes in the
circumstances of the Society. About the beginning
of the year, 1760, Mr. Lawson, the tutor, having
vacated his Fellowship, by being presented to the
Rectory of Swanscombe, in Kent, Mr. Elliston and
Mr. John Hey became joint tutors. Dr. Parris dying
soon after, and Mr. Elliston being chosen to preside
over the Society of Sidney, Mr. Martyn succeeded
to the tuition, as coadjutor to Mr. John Hey. Mr.
Byrch and Mr. Harvey having resigned their Fellowships, and Mr. Fearon having been presented to the
Living of Peasmarsh, in Sussex, the government of
the College now devolved *entirely* on the six Fellows,
who had been elected together only two years before!
This joint tuition continued for nearly 14 years; being
dissolved only by Mr. Martyn's marriage. Respecting this academical intercourse, a co-temporary, who
enjoyed Mr. Martyn's friendship to the close of his
life,—and who yet survives him,—bears the following
testimony:—" He told me one thing," says Mr. Richard Hey*, brother of the senior tutor, " which tends
to show the amiableness of his disposition; that ' there
never was the slightest disagreement between himself

* Letter to the Writer of this Memoir, dated Hertingfordbury,
30th Aug., 1828.

and his fellow-tutor, during the whole time of their acting together. This he mentioned, since the death of my brother John, in 1815. I was in some degree a witness of this amity; as I was Mathematical Lecturer of the College, from November, 1768, to the spring of 1775. Now, if we consider the great number of transactions which must have passed between the parties, in the course of fourteen years, this circumstance leaves a very pleasant impression of concord and amity."

Mr. Martyn was highly esteemed and respected by his co-temporaries in the University—most of whom he outlived. We shall have occasion, in the course of this Memoir, to advert to several of his Cambridge acquaintances; but we may here mention more particularly two of his friends at Emmanuel,—Mr. Hubbard*, his much respected tutor, who has been already

---

* Dr. HUBBARD was born in 1708, at Ipswich. He was educated at Catharine Hall, but removed to Emmanuel, in 1732. He was tutor to Mr. Hurd, afterwards Bishop of Worcester. In 1758, he was elected Registrar of the University. On the death of Dr. Richardson, he was unanimously elected Master of Emmanuel, 21st March, 1775; but, on account of his age and infirmities, with his wonted moderation and disinterestedness, he declined that honour." MSS. Brit. Mus. Cole, Vol. XLVI., p. 355. He died in 1778, and was buried in the College Chapel, where a tablet is erected to his memory. His epitaph concludes with the following high character:—" *Nullum hujusce sæculi virum, aut vivum majus coluerit, aut mortuum defleverit Academia.*" He had been Fellow forty-six, and Tutor, thirty-five years. See Nichols's Lit. Anec. of 18th Cent., Vol. II., p. 619, and Vol. IX., p. 507.

noticed, and Mr. Farmer,[a] his fellow-student, who afterwards became Master of the College. The following extract from a letter[b] written by him to Mr. Nichols, when almost an octagenarian, contains so lively a description of the persons and circumstances to which it alludes, that, while perusing it, we seem almost living in the midst of Mr. Martyn's academical circle :—

" I observe that you have been very large in your account of my intimate friend, Dr. Farmer. He was admitted of Emmanuel College, in October, 1752, and came into residence at the same time. Though he arrived within a few days after me, yet he was a year my junior ; because he had not been admitted before the commencement. We were near neighbours in Bungay-Court, and almost always together.

" I could have wished a fairer account of our most respectable tutor, and Dr. Farmer's firm friend, Mr. Hubbard. He was a tory, but not the least of a Jacobite ; nor was Dr. Richardson. They were both disciplinarians ; and considered *minutiæ*, perhaps with some reason, as the outworks of discipline. We see *now* [—this was written in 1813—] the consequence of their having been given up : the citadel

---

[a] Dr. FARMER was chosen Master of Emmanuel, 1775; he died in 1797. See a long Memoir of him in Nichols's Lit. Anec., Vol. II., pp. 618—649.

[b] Printed in Nichols's Lit. Anec., Vol. VIII., pp. 420—421.

has been stormed. Bickham, the junior tutor, was a bold man, and had been a bruiser when young. I do not think he was of any party. It is inaccurate to call him the *classical* tutor; for he gave us Lectures in Euclid. He did not want parts, but he was idle.

" Dr. Farmer's degree should not have been called *inconsiderable;* it was even reputable. Considering how idle he was, and how little inclination he had for mathematics, it showed the goodness of his parts. There was no contest between him and Sawbridge for the cup; Farmer had it of course, as senior in the Proctor's list. I was much oftener Curate of Swavesey than Dr. Farmer. Mr. Allenson, the Vicar, went every other year to see his relations in Yorkshire, and was absent 12 months. At these times, Dr. Farmer and I were his substitutes. I never recollect there being any *Methodists* in the parish.

---

* [This remark was occasioned by a note in the Literary Anecdotes, (Vol. VIII., p. 621,) taken from the Annual Necrology, to the following effect:—" Swavesey was at that time frequented by *Methodists;* occasioned by the Rev. H. Venn, then Rector of Yelling, in Huntingdonshire, and by the Rev. Mr. Berridge, then Vicar of Everton, Bedfordshire. Between these gentlemen and Farmer there existed no great cordiality; for Farmer was no friend to their doctrines; which appeared to him irrational and gloomy. Farmer was a greater adept in cracking a joke, than in unhinging a Calvinist's Creed, or in quieting a gloomy conscience."—Mr. Martyn objected, justly, to this ignorant and vulgar application of the term, *Methodist*, to those members of the Church of England, who had been deeply impressed by the preaching of the excellent men named in the preceding extract.]

Dr. Farmer was not famous as a preacher. His sermons were florid, and composed in haste; his enunciation was loud and hurried; his setting-off was so violent as to make nervous people start. As a proof of his hurrying, I heard him relate that, having been to preach at Huntingdon, and on his return riding over the bridge, he heard a man say to his companion, 'Aye, there he goes; if he rides as fast as he preaches he will soon be at Cambridge!' He was occasionally writing remarks on Shakspeare, from the first of his residing at Cambridge. I perfectly recollect his little *porte-feuille*, filled with scraps of paper of all sizes, in no order, which I occasionally attempted to arrange; and sometimes he would bring me some of his own writing to decypher, when he could not make it out himself. . . . He very justly writes *raptim*, or *calamo rapidissimo*, at the end of his letters; for he was always in a hurry. He suffered a disappointment in love very early in life. From his first coming to College, he always gave Miss Benskin, as a toast, and never could mention her name without evident feelings of the most ardent affection. We were then so intimate, that his joys and sorrows were poured into my bosom. After a lapse of almost sixty years, it is no wonder if I do not correctly remember how the connexion terminated; but I have some notion that at length she married another person; there being little prospect of the connexion with Dr. Farmer speedily taking place. But, as she was a Leicester girl, Mr. Nichols may perhaps know this

circumstance better than I do. This I am certain of,
that the disappointment affected his mind very
deeply, and was the source of his peculiarities. Of
his latter connexion with Miss Hatton, I cannot
speak with the same certainty, because at that time I
did not reside in the University ; and our intimacy
had ceased ; though we continued very good friends
to Dr. Farmer's death [in 1797] ; as, indeed, who
could be otherwise than friendly with so kind and
good-humoured a man as he was ? Dr. Colman was
likely to know the truth of the affair with Miss
Hatton. To the character given of him I make no
objection. The *Encyclopædia* [*Britannica*], Mr.
Isaac Reed, Mr. Dibdin, and Dr. Parr, have done
him justice. There is ' naught set down in malice ; '
nor is the truth concealed, nor even varnished.—I
still look back to him with great affection."

" THO. MARTYN."

We have now to consider Mr. Martyn in a different character from that of a fellow and tutor of
Sidney College ; and we must go back a few years
in his life, in order to trace the steps by which he
gradually attained to eminence in the elegant science
to which he became enthusiastically devoted about
this period. He had imbibed a taste for *Botany*, very
early, from his father ; and had pursued it with
ardour while he resided under his paternal roof at the
Hill House, on Streatham Common, in Surrey, in the
spring and summer of 1752. The Linnæan system,

had not then been introduced into England; although it had already been amply detailed in the writings of the illustrious Swede, by many successive publications, from the year 1732.   Mr. Martyn himself, indeed, assisted, as we shall soon see, to introduce it into this country.   " About the year 1750," he remarks, " I was a pupil in the school of our great country-man, Ray.   But the rich vein of knowledge, the pro-foundness and precision which I remarked every where in the *Philosophia Botanica*[a], (published in 1751,) withdrew me from my first master, and I became a decided convert to that system of Botany which has been since generally received.[b]"—" I had long before been acquainted with the *Systema Naturæ, Genera Plantarum*, and *Critica Botanica*, which Linnæus him-self had presented to my father. But, that inestimable work, the *Philosophia Botanica*, in 1751, and, above all, the *Species Plantarum*, in 1753, which first introduced specific names, made me a Linnæan completely[c]!"—" Being then (1753-6) engaged in academical studies, and afterwards (1756-9) in those of the profession I had determined to adopt, Botany

---

[a] Rousseau remarks, respecting this work, " c'est le livre *le plus philosophique* que j'ai vu de ma vie !"—" I never shall forget," writes Mr. Martyn, towards the close of his life, " how Dr. John Jebb was delighted with the Philosophia Botanica, when I put it into his hands, on his first attending my lectures. Jebb was a good philosopher, though a bad divine and politician."—Letter to Archdeacon Coxe, Feb. 10, 1808.

[b] Preface to the Language of Botany, 1807, third edit.

[c] Letter to Archdeacon Coxe, in 1809.

was rather the amusement of my leisure hours, than
my serious pursuit.*" In such modest terms does he
refer to the knowledge he at that time possessed of
the science in which he afterwards became so distinguished. It appears, however, by the following
letter from Mr. Pulteney[b], that his taste for Botany,

* Language of Botany, Preface. 1807. Third edition.

[b] The celebrated RICHARD PULTENEY, M. D., F. R. S., was born
Feb. 17, 1730. His friend and biographer, Dr. Maton, observes,
that, "the formation of that taste for which he became so distinguished, seems to have taken place in very early youth. Instead
of engaging in the boisterous and useless sports of his school-fel-
lows, in the hours of relaxation from business, he used to wander
in the fields with no companion but his Herbal, examining the
plants that grew in his path with the most lively curiosity." In
1764, he obtained his diploma at Edinburgh. He first settled at
Leicester; but, in 1765, he removed to Blandford; where he
practised as a physician till his death. In 1781, he published his
" *General View of the Writings of Linnæus.*" (This was republished by Dr. Maton, in 4to., 1805, with his Life of Dr. Pulteney.) In 1790 he sent out his " *Historical and Biographical
Sketches of the progress of Botany in England, from its origin
to the introduction of the Linnæan System.*" His contributions
to Nichols's History of Leicestershire, contain nearly 600 species of
the more rare plants found in the neighbourhood of Leicester, and
Loughborough, and in Charley Forest. He died Oct. 13, 1801,
and was buried at Langton, near Blandford. His widow erected
a tablet to his memory in Blandford Church; as he had " expressly
forbidden any eulogy to be inscribed on his monument," the simple
ornament of a PULTENÆA " delicately indicates the pursuits by
which he was distinguished."—The PULTENÆA, says Professor
Martyn, was " so named by Dr. J. E. Smith, in honour of my much
esteemed friend RICHARD PULTENEY, whose writings have so
essentially contributed to the introduction and establishment of

and his ardour in the pursuit of it, were by no means unknown to persons who were fully capable of appreciating his merits; and who eagerly cultivated his acquaintance from purely scientific motives :—

From Mr. Pulteney to the Rev. T. Martyn.

Leicester, Sept. 26, 1760.

Sir,

It is natural for those who are animated by the same genius, to seek and cultivate each other's acquaintance; and perhaps this principle is not more strongly felt in the lovers of any science than in that of *Botany*. This, I hope, will, in some measure, plead my excuse for the freedom of this address from one to whom you are entirely a stranger. If a further apology is necessary, I must add, that I was solicited to it by your friend and colleague, Mr. Farmer. I should be proud to add to my correspondents the son of a gentleman who has so long done honour to the science we both admire. I congratulate you on the prospect of having shortly a Botanic Garden at Cambridge, and am glad to find that so curious and useful a study is likely to meet with more countenance and encouragement, from our great people, than it has hitherto done ! . . . . .

Linnæan Botany in this country." Miller's Gard. Dict. by Martyn. —The *Pultenæa stipularis* was the species *first* discovered, and appropriated to this amiable Botanist.   See Smith's Bot. of New Holl., p. 35, t. 12, and Curtis's Bot. Mag., t. 475.

As Mr. Farmer tells me you sometimes, in your Botanic excursions, collect the English plants, I shall hold myself under obligation to you for duplicates of any in the subjoined catalogue. . . . .

There is now in London a quondam pupil of Linnæus; his name is *Solander*, which you will doubtless recollect to have seen in Linnæus's works. He is mentioned in the *Ratio Editionis* to the new *Systema Naturæ*. I have some expectation of seeing him at Leicester next summer, probably along with Mr. Hudson, and should rejoice in the pleasure of your company at that time; as Mr. Farmer has given me some reason to expect that you may be induced to take the trip with him next year. . . . .

I am, &c.,

R. PULTENEY,

This letter was the commencement of a correspondence between these two eminent Botanists, which continued for thirty-six years.

From the Rev. T. Martyn to Mr. Pulteney.

Cambridge, Oct. 9, 1760.

Sir,

I confess myself much obliged to you for beginning a correspondence, which I had long ardently wished for, and have often had thoughts of beginning myself; had I not been conscious how little I was likely to add to your knowledge, from my own small

stock; and from the little acquaintance and correspondence I have in the Botanical world. If I am
young in the science, it is, however, a great favourite
with me; and nothing rejoices me more than to cast
in my mite to promote it. I have already seen
enough of it to know that it is the distinguishing
mark of the professors of it, to communicate their
knowledge to one another, and to join their forces for
its enlargement. *So far* I boast myself to be *a Botanist;* and, therefore, shall esteem myself obliged to
you, if, at any time, you will employ me, either to
enlarge your collection of specimens, or to assist you
in any thing else; in return I shall make no conscience of troubling you in the same way, or with any
Botanical doubts or inquiries which may fall in my
way.

I have been pretty diligent this season in collecting
the vegetable inhabitants of this county; that, having nearly exhausted them, I might give myself up to
the foreign plants in our Botanic Garden; which,
indeed, has already *some* few in it; but *so* few that
they have taken me but little time to settle. The
foundation of the Green-house is laid, but the Garden
is not yet in any order; another summer we hope will
produce something.

I am sorry it is not in my power to send you *all*
the plants in your list. The *Antirrhinum minimum*ª,

* [*A. Orontium*, Lesser Snap-Dragon of Eng. Flora.]

Aphaca[a], *Calamintha vulgaris*[b], *Cyperus inodorus*[c], *Jacobæa montana*[d], *Linaria adulterina*[e], *Melampyrum cristatum*[f], *fl. purp.*, are all I now have. The *Alysson Germ. echioides*[g], is said, by Mr. Ray, to be lost near Newmarket, and has not been found since, either by my father or myself. . . . . I have sent you a few which you did not mention; if any of them are not already inhabitants of your Hortus Siccus, I shall be glad. The *Bupleurum minimum*[h], I found, in August last, by the side of Hinton Moor, near Cambridge; the *Lathyrus major latifolius*[i], in Kingston Wood; the *Caryophyllus minor repens*[k], near Hildersham; *Ros Solis fol. oblongo*[l], on Hinton Moor; and the *Scrophularia vernalis*[m], at Mitcham, in Surrey, where it was

[a] [*Lathyrus Aphaca*, Yellow Vetchling.]
[b] [*Thymus Calamintha*, Common Calamint.]
[c] [*Cladium Mariscus*, Prickly Twig-Rush.]
[d] [*Senecio sylvaticus*, Mountain Groundsel.]
[e] [*Antirrhinum spurium*, Round-leaved Snap-Dragon.]
[f] [Crested Cow-wheat.]
[g] [*Asperugo procumbens*, German Madwort. It was again found near the church at Newmarket, by the late Mr. Relhan.]
[h] [*B. tenuissimum*, Slender Hare's Ear.]
[i] [Broad-leaved Everlasting-Pea.]
[k] [*Dianthus deltoides*, Maiden Pink.]
[l] [*Drosera longifolia*, Long-leaved Sun-Dew]
[m] [*Yellow Figwort.* Mr. Smith, a surgeon, of Croydon, had the merit of adding this rare plant to the English Flora. He found it in 1759, on " the left of the road from Mitcham to Morden." His letter, dated April 26, 1759, communicating the discovery to Professor John Martyn, was given by Professor Thomas Martyn to Sir Joseph Banks, and is now among the Banksian MSS., in the

shown me last year by Mr. Smith, of Croydon; who I suppose was the first who found this plant growing wild in England.

I cannot but say I was sorry, when I heard that Hill was about a *Flora Britannica*: it is, indeed, a most vile performance; he has not taken the pains so much as to alter the errors of the press; nor do I find any thing new, except the addition of new places, which he has from you or Dr. Watson.

I have not the pleasure of any acquaintance with Mr. Hudson, but have often heard of his merit, and of his abilities for the work[*] which he has in hand, and which I expect with some impatience. . . . . I am informed he intends a visit to Cambridge next summer; if so, I could wish to see him; not only that I might show him the civilities of the University, but because, perhaps, he might like to look over my collection of English Plants; which, by my father's diligence, are not far from complete. I wish, too, Mr. Solander would make a visit here, that I might have the pleasure of conversing with a pupil of Linnæus, of whom I have heard a very good character. I had rather, indeed, meet him and Mr. Hudson with you at Leicester; and, if I can contrive it, I certainly *will*. . . . . . I will now give you a list of such plants

British Museum. None of the British Floras name the real discoverer; all quoting *Hudson* for the Surrey habitat. Professor T. Martyn quotes it from " Mitcham Common Field, in the Lane to Merton."

[*] *Flora Anglica*, published in 1762.

as I want. . . . . . In return, I hope for a longer list from you, against the next year, which I intend should be the last of my peregrinations in this county. . . . . Mr. Farmer, whom I have the pleasure to place among the first of my friends, desires to be remembered to you.

I am, &c.,

Tho. Martyn.

From the Rev. T. Martyn to Mr. Pulteney.

Cambridge, April 2, 1761.

Sir,

I am almost ashamed to see, by the date [Nov. 8, 1760] of your last obliging letter, how long it is since I ought to have written to you. One reason for my delay was, that I might send you the produce of my winter excursions. . . . . You will see that the Catalogue of Cambridge Mosses is somewhat increased. . . . . It may, perhaps, be some satisfaction to you that our Garden begins to flourish. Shrubs and trees in abundance are already planted; plenty of seeds, both tender and hardy, are sown; a stove is building; and stone is preparing to raise the superstructure of a Green-house, on the foundation which was laid last year. All this, I hope, will increase the number of Botanists among us. Indeed, we already begin to grow considerable; for I never had more than *one* companion before this spring, but now I have *three;* and expect soon to have two or three more converts.

. . . . Some time next spring, I will take the liberty of sending you an account of all the rarer English plants which are not in my Hortus Siccus; hoping that you will do the same; that we may use our mutual endeavours to complete each others collections. . . . . I found the *Ceratophyllum demersum* β[a] last year for the first time; whether it is the same with the other or not, I cannot pretend to say. . . . Surely there is some blundering about the *Rumex aquaticus*, if by that our *Common Water Dock* be meant; for who could ever discover, in *that*, "heart-shaped" leaves?—which it ought to have according to Linnæus's name[b]. . . . . .

I am, &c.,

THO. MARTYN.

From Mr. Pulteney to the Rev. T. Martyn.

Leicester, 9 Apr., 1761.

Sir,

. . . . It is a great pleasure to me to hear that your Garden is likely to flourish; but particularly

---

[a] [*C. submersum*, or Unarmed Hornwort: it is now ascertained to be distinct from the common species, *C. demersum.*]

[b] [Mr. Martyn was quite correct in his suspicion. The English plant is the *Rumex Hydrolapathum*, or Great Water-Dock, of Hudson; which is now justly distinguished from the true *R. aquaticus* of Linnæus, a Swedish plant. Linnæus was led into an error, by the near resemblance between these plants. The mistake was followed by Dr. Smith, in the *Flora Britannica*; but he has fully corrected it in his admirable *English Flora.*]

that you begin to enlarge your circle of Botanic friends upon the spot. I shall be proud to hear that the genius of Botany raises her head in so conspicuous a situation : you have my most ardent wishes for her welfare; nor do I doubt but she will meet with all the encouragement that you can lend her, and, in a little time, flourish happily under your auspices. . . .

RICHARD PULTENEY.

From the Rev. T. Martyn to Mr. Pulteney.

Cambridge, April 19, 1761.

Sir,

I am glad to embrace the first opportunity that offers, to answer your obliging letter. Mr. Ashby*, whom I have lately become acquainted with, from the similarity of our pursuits, and who is extremely diligent and ingenious in natural inquiries, does me the favour of conveying this to you.

You will find, in the parcel, the mosses which you desired me to send, &c. &c.; . . . . and a proliferous head of the *Juncus articulatus*ᵇ, so formed by the plant

* [The Rev. G. Ashby was born Dec. 5, 1724, and died June 12, 1808. He was of St. John's College, Cambridge, and a man of considerable talents. Mr. Martyn enjoyed his friendship from 1761, till his death in 1808. He was presented by his College to the Living of Barrow, in Suffolk. There is a short account of him in Nichols's Illustrations of Lit. Hist. of eighteenth Century, Vol. IV., p. 887.]

ᵇ [*J. uliginosus*, or Little Bulbous Rush, of the English Flora ; the variety marked β by Dr. Smith, with viviparous flowers.]

bending down to the water when the seed was ripe: the reason of my sending you this, is, because I apprehend that the specimen *you* sent was formed in the same manner, and is no otherwise different from the *Juncus bufonius*[a].

Business will detain me here till the 8th or 10th of next month; after which I shall make my father a visit for some time, at Streatham, where I shall be for the most part, though now and then at Chelsea. Mr. Charles Miller and I intend to *walk* it, for the sake of *simpling*. . . . .

Tho. Martyn.

From the Rev. T. Martyn to Mr. Pulteney.

Cambridge, Sept. 10, 1761.

Sir,

When you did me the favour of a letter, (July 25,) I had just reached Cambridge, where I have been ever since. Having nearly exhausted the plants of this county, my walks have been but few this summer; especially as the vicinity of the Botanic Garden is a decoy to an idle man. You have one plant from Gamlingay inclosed, and there are two or three more in reserve for you; that you have not more specimens is, in some measure, your own fault,

[a] [Or Toad-Rush.—Mr. Pulteney's specimen came from a tarn, in Yorkshire. " I take it," he writes, " for the *Graminifolia lacustris prolifera, seu plantulis quasi novis hinc inde a cauticulis succrescentibus* Raii Syn. II. 183. Syn. III. 134."]

in not letting me know what will be most acceptable
to you; and the rarer plants of one county being
often common in another, it is impossible to know
which plants are most desired.

I hear that the Prolegomena of the *Flora Danica*
is arrived in England; but I have not yet seen it.
No doubt it will be a very pompous work, as the
King of Denmark has given £6000 towards the encouragement of it. It will come tolerably easy as to
price; for they publish sixty plates every year for two
guineas. The Prolegomena will probably inform us
of their design. . . .

THO. MARTYN.

A long list of Cambridgeshire Lichens, Mosses, and
other Cryptogamous plants, discovered by Mr. Martyn
in his rambles, forms the postscript of this letter. It
appears, indeed, from the correspondence between
Mr. Pulteney and himself, at this period, (the whole
of which is still preserved), and from their frequently
exchanging rare plants, that he had at this time given
a very zealous attention to his Botanical studies. His
scientific merit and diligence soon met with a suitable reward. At the close of the year 1761, Professor Martyn, the elder, having signified to the Vice-
Chancellor his wish to resign the Botanical Chair, on
account of his increasing infirmities, Mr. Thomas
Martyn was unanimously elected his father's successor, on the 2nd of February, 1762. " I congratulate you,"—writes Dr. Pulteney, on the 23d of April,

of that year,—" upon your election to the Professorship of Botany; I hope the science will flourish
under your auspices; and I most cordially wish that
this distinction may be attended with all the consequences that can afford you pleasure in a science
which you have so much at heart." Few circumstances, indeed, could have been so gratifying to him
as this appointment;—his learned father had occupied the Botanical Chair for twenty-nine years; Mr.
Thomas Martyn filled it for sixty-three years: the
Professorship was thus held with credit by two successive generations of the same family, for nearly a
century!

The Institution of a Botanic Garden, which took
place a little before his election, (about 1761,) was
particularly favorable to Professor Martyn's pursuits.
This had always been a desideratum at Cambridge.
So long ago as 1696, the ground for a " Physic
Garden" had been measured, and the plan drawn*:
but, through some unknown impediment, the scheme
failed.   Professor Bradley made large, but hollow
promises on the same subject, in 1724, with the mere
view, (it should seem,) of obtaining the Botanical
Chair;—he publicly repeated them in his lectures, in
1729; but nothing was done.   In 1731, there appeared more hope—for many conferences were held
between the Vice-Chancellor, Professor John Martyn,

---

* See Cole's MSS., Vol. XXXIII., p. 26, and Cole's Athenæ
Cantab. (MSS.) Vol. III., p. 312.

and Mr. Phillip Miller, of the Chelsea Garden, respecting the estate of a Mr. Brownell, of Willingham, which was once intended to be devoted to the establishment of a Botanic Garden, at Cambridge; but this estate was diverted into another channel. At length the plan was happily effected, through the liberality of Dr. Walker, the Vice-Master of Trinity College, who gave an estate to trustees for that purpose. The ground selected for the Botanic Garden was the site of the Monastery of the Austin Friars, and was purchased by Dr. Walker for £1600, in 1761. Early in the following year, he appointed Professor Thomas Martyn the first Reader, and Mr. Charles Miller* the first Curator, in his new establishment.

This year he was elected Junior Proctor of the University; instead of the person who had been nominated by Magdalen College, but who became incapable of executing the office a few days after his election.

* CHARLES MILLER was born in 1739. He was the youngest son of Philip Miller, the celebrated Curator of the Physic Garden at Chelsea. He executed the trust of Curator of the Cambridge Garden, with great satisfaction to the University, for several years; but resigned it in 1770, and went to the East Indies; whence he returned, and was resident in London, where he died, Oct. 6, 1817. In Vol. LXVIII., p. 160, of the Phil. Trans., is a very interesting account of the Island of Sumatra, collected from letters to his friends, but published without his knowledge. Several valuable letters, addressed by him to Professor Martyn, between 1771 and 1774, are among the Banksian MSS. in the British Museum.

From Professor Martyn to Mr. Pulteney.

Sid. Coll., Dec. 27, 1762.

Sir,

It is with shame that I acknowledge myself to have been a very long time in your debt. . . . The seeds of the *Campanula patula**, which you were so kind as to send, I have given to Mr. Miller, and hope to see them adorn our Garden next summer . .

The first volume of Linnæus's *Species Plantarum* is printed off, and is increased in bulk, I am informed, one fourth : I suppose we shall be longer than we could wish before we see it, as we must wait for the second.

I have just got the first Fasciculus of *Danish Plants*, consisting of sixty plates.   The engraving is perfectly neat; but Mr. Miller tells me the painted ones are better done than any thing of the kind we ever had; I have not had a sight of these yet, but am promised it soon.

You see how regal Dr. Hill is, in *giving away* the editions of his late work! which is now published both in 4to. and 8vo.

You have, doubtless, by this time studied through Mr. Hudson's *Flora Anglica,* which is certainly a very useful book.   But I could have been glad to have seen *all* the English Botanists called on to concur in such a work; by which means some additions might

---

* [Spreading Bell-Flower.]

have been made, and many inaccuracies avoided; but an undertaking of that kind, will, perhaps, best flourish under *Royal*[a] protection.

A multiplicity of business has hindered me from making any Botanical excursions this year; another, I hope, will turn out better.

T. MARTYN.

From Mr. Pulteney to the Rev. Professor Martyn.

Leicester, Feb. 5, 1763.

Sir,

. . . In a letter I lately had from Dr. Hope, Professor of Botany at Edinburgh, he informs me that they have discovered the *Betula nana*[b], and the *Arbutus Uva-Ursi*[c], to be natives of Scotland. Doubtless, if the Caledonian Alps were well searched, we should have many more of those Alpine plants found which Linnæus mentions in the *Flora Lapponica*, and which we are strangers to, yet, as natives of Britain. . . .

R. PULTENEY.

In the spring of 1763, Professor Martyn read his first Course of Lectures, to fifty pupils, besides about as many more occasional hearers. He showed his disinterested love for science, by admitting to these Lectures, *without the customary fee*, every person

---

[a] [It has been completed, however, and read by *private* exertions alone.]

[b] [Dwarf Birch.]   [c] [Red Bear-Berry.]

who would subscribe £10. to the new Garden, which
was by no means adequately provided for by Dr.
Walker's grant, and the subscriptions towards which
had advanced but slowly.   His pupils were delighted
at recovering a science which had been almost dor-
mant twenty-eight years. Dr. Heberden, had, indeed,
about the year 1748, given a Course of experiments
on the *Medicinal* plants of Cambridgeshire; but the
want of a Garden rendered it impossible for him to
proceed with effect; and his profession soon called
him away from the University.   Professor Martyn
had now the honour of introducing the Linnæan
system to the University, in the *first* Course of Lectures ever read in England, founded upon the method
of the illustrious Swedish Naturalist.   The æra of
the introduction of Linnæan Botany into England,
was, indeed, somewhat earlier, as regards *publications*
on the subject. Stillingfleet first called the notice of
our countrymen to the system of Linnæus, by the
publication of his " Miscellaneous Tracts, on Natural
History," in 1757.   Lee, in 1760, sent out his " Introduction to Botany," in which the principles of the
great Swedish Teacher were fully explained to the
English student.   Dr. Hill's " Flora Britannica," published the same year, adopted the classes and generic
characters of Linnæus, but not his nomenclature.
In 1762, Hudson, published his " Flora Anglica;" a
work which marks the complete establishment of the
Linnæan nomenclature, as well as of the system; and
which became the universal manual of British Bo-

tanists. All these works had been printed during
the five years which preceded that in which Professor Martyn delivered his first Course of Lectures
at Cambridge ; but it was no small honour to be
the first public advocate, and the earliest promulgator
of the Linnæan system of Botany, in an English
University.   " The Linnæan language and system
were at that time entirely new to the University, and
very little known, or attended to, in other parts of
the kingdom, except at Edinburgh, by the laudable
efforts of the late Dr. Hope.ᵃ"   That distinguished
individual began his Botanical Lectures, in the Caledonian Metropolis, cotemporaneously with Professor
Martyn in Cambridge ; who always took particular
pleasure in associating Dr. Hope's labours with his
own, in his scientific recollections.

From Mr. Pulteney to the Rev. Professor Martyn.

Leicester, May 1, 1763.

My dear Sir,

        " I wish you all possible success and en-
couragement in the prosecution of your Lectures,
which I hear you have lately begun, and as I presume you will put the science of Botany into a new
dress, by the method and improvements of the illustrious Swede, I doubt not but you will add even new
dignity, if that can be, to the Chair which your
worthy predecessor so eminently filled. · On the

ᵃ Martyn's Lang. of Botany, Preface.

second day of April, Dr. Hope began *his* Botanic
Course; in the introductory Lecture of which, he
exerted himself much to excite the students to lend
him their assistance in collecting plants, as he conceived a design of publishing a *Flora Scotica*. This
will certainly bring us many new plants to light, if the
Highlands be duly scrutinized. He rewarded with a
gold medal the gentleman who made the best collection last year, and thanked him publicly before his
class. The specimens were almost as green as when
growing, and the colours of the flowers were uncom-
monly well preserved.

RICHARD PULTENEY.

He now published the fruits of his researches, during his Botanical rambles in Cambridgeshire, under
the following title : *" Plantæ Cantabrigienses ; or, A
Catalogue of the Plants which grow wild in the County
of Cambridge, disposed according to the system of Linnæus.—Herbationes Cantabrigienses ; or, Directions to
the places where they may be found, comprehended in
Botanical Excursions. 13.—To which are added, Lists of
the more rare plants growing in many parts of England and Wales. By* THOMAS MARTYN, M. A., *Fellow
of Sidney College, and Professor of Botany, in Cambridge. London, 1763,"* 8vo. pp. xiii. and 114. This
volume is dedicated " to John Martyn, F. R. S., and
late Professor of Botany in the University of Cambridge, in gratitude for his eminent paternal affection,
and that readiness which he has always manifested to

promote the cause of Botany in general, and the Botanical studies in particular of his dutiful son." As this little work is intimately connected with the formation of the CAMBRIDGESHIRE FLORA, we shall enter upon a more particular account of the various contributions of eminent Botanists, by which the catalogue of plants, growing spontaneously in that county, was from time to time enlarged, till it at length became the most complete *local* Flora of this kingdom.

The illustrious Ray laid the foundation of the FLORA CANTABRIGIENSIS, by the publication of his "Catalogus Plantarum circa Cantabrigiam nascentium, in quo exhibentur quotquot hactenus inventæ sunt, quæ vel sponte proveniunt, vel in agris seruntur ; unà cum Synonymis selectioribus, locis natalibus, et observationibus quibusdam oppidò raris. Cantab. 1660." 12mo. pp, 182. With this was connected, by a separate title, " Index Plantarum Agri Cantabrigiensis, in quo nomina Anglica Latinis præponuntur, ordine alphabetico, in gratiam tyronum. Cantab. 1660." 12mo. pp. 103. Ray acknowledges the assistance of Mr. Nid, Fellow of Trinity, and of Mr. Francis Willoughby, and Mr. Peter Courthope.—An " Appendix" was published in 1663 ; and " Appendix altera" in 1685. Ray's Alphabetical Catalogue, with its two Appendices, (to the last of which, Mr. Dent, a medical gentleman of Cambridge, contributed 40 plants,) contains 726 species.

Professor John Martyn, six years before his election to the Botanical Chair, republished the Cata-

logues of Ray, for the use of his pupils; arranging
the plants according to the *system* of that eminent
naturalist, instead of following the *alphabetical* plan.
His little volume, entitled " Methodus Plantarum
circa Cantabrigiam nascentium. Lond. 1727," 12mo.
pp. viii. and 132,—has been already* noticed. This
work contains no more plants than Ray's. The author began to print a new edition in 1729, containing
many new plants; but it was never completed.

The Cambridgeshire Flora was left in this state,
till 1763, when Professor Thomas Martyn published
his " *Plantæ Cantabrigienses*, &c.," and " *Herbationes
Cantabrigienses*, &c.," both in one volume, of which
the full title has been given a few paragraphs back.
In this little work, 181 new plants were added to the
preceding Catalogues; the whole number being thus
advanced to 907. In the "*Plantæ Cantabrigienses*," the
arrangement is according to the Linnæan classes, but
in three parallel columns; of which the first exhibits
the generic and trivial names of Linnæus,—the second gives the name, with reference to the page in
Professor John Martyn's "Methodus Plantarum;"—
the third shows the name in Ray's alphabetical " Catalogus Plantarum." This arrangement was particularly valuable, and indeed almost necessary, at a time
when the Cambridge plants were better known by the
synonymous phrases of Ray, than by their modern
Linnæan names; although it is now superseded, except

---

* See above, p. 34.

for critical reference. The "*Herbationes Cantabrigienses*," points out, in thirteen excursions, all the plants which may be met with; the habitats being noticed in a general manner in the text, and more circumstantially in the notes with reference to remarkable plants. This is an excellent plan; a republication of the "*Herbationes*," adapted to the present complete state of the Cambridge Flora, would form an useful manual for Botanical students. The Appendix mentions many rare plants, ranged under the several counties; the author frankly acknowledges the defectiveness of this part of his volume; and, indeed, it has been quite superseded by later writers on the same plan.

The same year in which Professor Martyn published his "*Plantæ Cantabrigienses*," there appeared a smaller work on the same subject, by Lyons[*]: "Israelis Lyons, jun., Fasciculus Plantarum circa Cantabrigiam nascentium, quæ post Rajum observatæ fuêre. Lond. 1763." 8vo. pp. 56. This contains a list of plants not known to Ray, found between 1727 and

[*] Israel Lyons, the son of a Polish Jew, (who was a silversmith and teacher of Hebrew at Cambridge,) was born in 1739. He displayed great talents when a young man, and was much noticed by Dr. Smith, then Master of Trinity College. He began the study of Botany when 16 years old; and had a remarkable talent for remembering not only the Linnæan names of plants, but also the perplexed and endless *synonyms* which had so greatly confused the science. Sir Joseph Banks was his pupil, and sent him to Oxford, about 1762—3, to read lectures; which he did with great credit, to about 60 pupils. He died in London, 1775.

1763, by Professor John Martyn, Charles Miller, Lyons, and others. Professor Martyn, in a letter written to Mr. Pulteney, this year, expresses the following opinion of this little work.—" It contains about 30 plants, not remarked in my book. Excepting, however, the Cryptogamia, the chief value of this Fasciculus is, some good observations and descriptions, for the accuracy of which I will venture to answer. You will see, by the author's preface, that he intends a ' FLORA CANTABRIGIENSIS ;' which he never signified to me, or else, perhaps, I had spared myself the trouble of printing my little pamphlet." (Oct. 15, 1763.) Haller says that the whole number of plants noticed in this *Fasciculus,* is 106. (Haller, Bib. Botan., Tom. II., p. 531.)

Professor Thomas Martyn continued his researches for Cambridgeshire plants, with great diligence, during thirteen years after the publication of his *Plantæ Cantabrigienses;* when he left the county altogether, to reside upon a distant living, in 1776. In 1777, he determined to put to press an enlarged edition of his former work; for which he had long since made considerable preparations. "I am now very busy," he observes, (addressing Dr. Pulteney, Aug. 6, 1777,) "on a FLORA CANTABRIGIENSIS, having laid it aside ten years, which is one more than Horace prescribes." He had even proceeded so far as to commit his MSS. to a friend resident in the University, the Rev. Michael Tyson, Fellow of Corpus, who had engaged to

correct the sheets*. The work was to have appeared
the following spring; but was stopped for further improvements, by references to the second edition of
Hudson's " Flora Anglica," which was on the eve of
publication : it was again postponed on account of
Professor Martyn's absence on the Continent, from
1778 to 1780,—and was never resumed.

His labours, however, were not lost to the public ;
though they obtained no further credit for *himself*.
With that love for science, and that liberality of mind
which so particularly marked his character, in 1783,
he generously transferred the whole of his MSS. materials to the Rev. Richard Relhan, Fellow of King's
College ; who now designed to bring out the long desired " Flora Cantabrigiensis," which the death of Mr.
Lyons, and the absence of the Botanical Professor
from the county, had left to be completed by other
hands. Mr. Relhan was too sincerely grateful to
wish to conceal his great obligations ; but Professor
Martyn was too modest and disinterested to allow the
full extent of his assistance to be publicly avowed ;
he therefore prohibited any thing beyond a *general*
complimentary acknowledgment in the Preface to the
intended work. The following passages from letters
(preserved in the Banksian MSS., in the British

* See Mr. Tyson's Letter to Mr. Gough, in Nichols's Lit. Anec.,
Vol. VIII., p. 628.

Museum,) addressed by Mr. Relhan to the Professor, are too creditable to both parties to be suppressed.

"12th August, 1783.—How is it possible that I can ever make any return for your inestimable present. You tell me much remains to be done ; for once suffer me to dissent from you, and to add that I am perfectly convinced that the excellencies of *Flora Cantabrigiensis* will be derived from your assistance. I beg leave to avow my obligations in my Dedication and Preface ; I cannot tell the world half what I feel. When I think of my undertaking, I tremble ; when I see your MSS. I take courage."

"Nov. 26, 1783.—I thank you for your peculiar delicacy with respect to me. I am anxious to tell the world how warmly I feel my obligations. I have seriously begun my difficult task ; and though I shall get very well through that part of the Flora which will be chiefly yours, yet I shall look forward to *Cryptogamia* with terror."

"Dec. 9, 1783.—Dickson, of Covent Garden, is a most wonderful fellow ; he was quite uneducated ; but, by his own industry, has acquired an accurate knowledge of the Cryptogamian plants. His zeal for science must have been amazing ; and he communicates the result of his labours with a generosity which I have experienced from only Mr. Martyn, and a very, very few."

This valuable work, which embodied the discoveries of Ray, of the two Martyns, of Lyons, and of the indefatigable author himself, appeared in 1785, under the following title.—" Richardi Relhan, A. M., Collegii Regalis Capellani, Flora Cantabrigiensis, exhibens Plantas Agro Cantabrigiensi indigenas, secundum systema sexuale digestas. Cantab. 1785." 8vo. pp. 490. An Appendix was published in 1786, pp. 38; a second Appendix 1788, pp. 36; and a third Appendix in 1793, pp. 44. The original volume, which was dedicated to Professor Martyn, together with its three Appendices, added no fewer than 383 new plants, discovered since the year 1763, (the greater additions being in the class Cryptogamia,) and advanced the whole number to 1212 species. The 2d edition was published in 1802, (pp.); it added 132 new plants, and brought the total to 1344 species. The 3d edition came out in 1820; it supplied 75 plants which had eluded former research, and advanced the number of species to 1419, natives of Cambridgeshire, which exceeds that of Sibthorpe's Oxford Catalogue by 144 plants. It has thus been ascertained that the county of Cambridge contains *more than one half* of the vegetable productions of Great Britain! The greatest praise is due to the late Mr. Relhan, for his indefatigable industry in bringing the Cambridgeshire Flora* to such a degree

---

* As soon as he had brought the Cambridgeshire Flora to a conclusion, Mr. Relhan proposed (in 1785) to undertake a *Flora Anglica;* but he did not meet with sufficient encouragement to prosecute his design. Mr. Relhan died soon after the publication of the 3d edition of his *Flora Cantabrigiensis;* viz. March 21, 1823.

of completeness; for though he had several diligent predecessors in this rich Botanical field, in the phæ-nogamous classes, he certainly gleaned indefatigably even in that department; and he reaped a harvest almost entirely his *own*, among the cryptogamous plants. This merit was cheerfully ascribed to him by Professor Martyn; who, in acknowledging the receipt of the second, and afterwards the third edition of Mr. Relhan's work, thus writes, in 1804 and in 1820,—" I cannot but admire the great pains and diligence you have used, in more than doubling the Catalogue of the indefatigable Ray."—" Your Flora keeps up the reputation of our County and University, and preserves the decided eminence of our local Flora above all others. *Cryptogamia* is nearly your own, without competition; nothing, in a manner, having been done in that class by any of us, even the great Ray himself."

We must now return, from this digression, to the summer of 1763, when Professor Martyn was finishing his first course of Botanical Lectures at Cambridge. He continued his Lectures without interruption, (except during 1779 and 1780, when he was abroad, and 1785, when his lecture-room was pulled down,) for the long period of *thirty-four years,* i. e. till 1796. When he had read Lectures in the science, which was his favourite study, for a few years, he found it necessary to add the other branches of Natural History, Animals, and Fossils; *Botany* not being then sufficiently popular to keep together a class on that single subject!

From Professor Martyn to Mr. Pulteney.

Cambridge, June 30, 1763.

Sir,

You will be glad to hear that the Botanic Lectures have been well attended. I have had 50 pupils; and though, probably, but a small proportion of this number will attend to much purpose, yet, upon the whole, if we can keep up our Garden, the science will certainly flourish among us. You may be sure that I teach the system of Linnæus. It rejoices me greatly to hear that Botany is so likely to flourish at Edinburgh, under Dr. Hope, especially as Scotland has been hitherto so ill searched. His designs, and the hope of having the pleasure to see *you*, had almost prevailed upon me to make the northern tour of the island; but I have now changed my resolution, and intend to set out, the beginning of next week, for Holland, from whence I shall proceed through Flanders into France. . . I greatly suspect that Hudson is mistaken, frequently, with regard to *Grasses*. . . The *Synonyms* of plants are so much confused, that it would be a Herculean labour to extricate them; and yet, I often wish that Dr. Sherard's collection was here, instead of at Oxford; because then, with the assistance of some ingenious friends, it might be hoped that, some time or other, this great *desideratum* of Botany might be completed ; and, were it not for my necessary avocations,* there is nothing I should so much desire to labour at, notwithstanding I am sensible of my own deficiency, and the greatness, as well as dryness, of the task.

T. MARTYN.

He spent the months of July and August on the Continent, travelling with his friend Mr. Lobb, Fellow of Peter House. On his return, he addressed the following letter to his esteemed Botanical correspondent, Mr. Pulteney:—

Cambridge, Oct. 15, 1763.

Sir,

    I could not let Mr. Ashby come into Leicestershire, without writing a line to you, though I have not much to say. I have been in England about six weeks ; after a very pleasant tour through Holland and Flanders, and three weeks stay at Paris. You will suppose that I did not neglect seeing the Botanic Gardens in my way. Those of Holland are very rich indeed ; particularly that of Leyden, in which Professor Van Royen did me the favour to show me many plants which were entirely new, and many more which were new *to me*. The Paris Garden is in a sad disorder, and in wretched condition,—full of weeds, and ill furnished with plants ;

---

* [He alludes to his College Lectures.]

K

which had a very bad appearance, after seeing the
extreme neatness of the Dutch Gardens, and the immense variety of plants which they contain. . . .

THO. MARTYN.

In 1764, he prepared for assembling his class,
by printing "*Heads of a Course of Lectures in Botany*:" it was not published, but given to his pupils :
a great part of the impression was burnt. On the
10th of October, he was appointed Junior Proctor of
the University ; an office which he had filled in 1762.
He now published three papers in the " Museum Rusticum ;" viz. :—

1. "*A list of the most remarkable Weeds which infest
the arable and grass lands in England, ranged according to their duration ; with the times of their flowering,
and hints on the best means of extirpating some of them.*"
Museum Rusticum, Vol. V., for 1765, No. LVI., pp.
301—308. Under the signature P. B. C. [*Professor
Botanices Cantabrigiensis.*]

2. "*Descriptions of those Weeds, enumerated in
Vol. V., No. LVI., that are annual, and not universally
known.*" Museum Rusticum, Vol. VI., for 1766, No.
XXVIII., pp. 194—199. *With fifteen figures.* Under
the signature P. B. C.

3. "*Descriptions of the biennial and perennial Weeds
enumerated in Vol. V., No. LVI.*" Museum Rusticum,
Vol. VI., for 1766, No. LXIV., pp. 441—453.

On the 11th of June, 1766, he proceeded to the
Degree of B. D.—The same year, he published ano-

nymously, "*The English Connoisseur: containing an account of whatever is curious in Painting, Sculpture, &c., in the Palaces and Seats of the Nobility and principal Gentry of England, both in Town and Country. London, 1766.*" 2 Vols., 12mo., pp. x. and 400. Granger observes that, "the mistakes in this book are not owing to any want of care and industry in the ingenious compiler, but to the inaccuracy of some of the owners of the pictures mentioned in the work."—Mr. Martyn makes some good remarks upon the difference between the Italian and Flemish Schools. He observes, that that taste is to be commended which prefers "the greatness of design and composition, in which the Italian masters are so well known to excel, before the gaudy Flemish colouring; or '*the drudging mimicry of nature's most uncomely coarseness*'[a], upon which the Dutch so much value themselves. To deny these their proper share of merit, or to refuse them a place in a collection, would be ridiculous; but, surely, to set them in competition with Italian sublimity, is much more so." (Pref., p. iv.)—The author digresses to a minute account of the Leasowes, and of Hagley Park; and, apologizing for having introduced matter somewhat foreign to the plan of his work, he adds,—"there is at least as much room for exercising the great arts of design and composition in laying out a *Garden*, as in executing a good *Painting*" (Pref., p. v.); —a passage in which we trace the strong inclination

[a] "Ædes Walpolianæ, Introd."

of his thoughts to his favourite study, even when writing on a very different subject!

In 1766, he writes as follows, to his friend Dr. Pulteney; somewhat discouraged at the little zeal for Botanical studies which he had been able to awaken in his University :—

Cambridge, May 31, 1766.

Dear Sir,

I thank you for the pleasure you gave me in the perusal of your elegant Latin Thesis*, and sincerely hope that it will be productive of all the happy consequences which you so well deserve. . . . You are now so far removed[b], that I have little hope of accomplishing a meeting which I have so long desired; and must be contented with the continuance of your correspondence. . . . . .

I have almost ended my Course of Lectures for this year. My pupils are but few in number; and there are fewer still who give any attention to the science. I hope, however, by perseverance, to bring it more into repute among us. . . . . The Garden gets on very well in point of plants, under the direction of Mr. Miller; but our income is still very scanty, so that we cannot finish our Green-house, much less build stoves: indeed, we are obliged to use a degree

* [Read at Edinburgh, for the Degree of M. D.]
[b] [Dr. Pulteney left Leicester, in 1765, to practise as a Physician at Blandford, in Dorsetshire.]

of frugality not very consistent with the dignity of an University, or the usefulness of the design; but we keep it on foot for better times !

I have the pleasure of hearing now and then from Dr. Hope, and he is so kind as to send me some seeds. . . . .

Mr. Banks, a gentleman of fortune, lately of Oxford, and extremely fond of natural history, in all its branches, is going on a very extraordinary expedition to the Esquimaux Indians, in search of the productions of nature. It seems a bad country for the purpose; but he is determined to that, because a friend of his is sent over by the Government, in a man-of-war, to transact some business with the people of those parts, and is accompanied by two others of his acquaintance. . . . .

THO. MARTYN.

Dr. Pulteney complained, in many of his letters addressed to Professor Martyn, of the want of scientific society in his new situation; and feelingly expressed his regret that, while his friend enjoyed the privilege of living in the very seat of science, *he* was almost cut off from commerce with literary men. To these regrets, the following letter refers :—

Chelsea, Feb. 18, 1767.

Dear Sir,

I am sorry to find that your establishment at Blandford is likely to be a bar to your Botanical inquiries. I had flattered myself, from your known abilities and love of the science, with hopes of considerable improvements from your hand; but you cannot, I am sure, *entirely* forsake *the flowery path;* and I know not whether I ought to wish you so much business as will leave you no leisure for your favourite pursuits.

As you express a desire of knowing how our literary world goes on, I will trouble you with the little news I have picked up; begging your excuse for what you have heard already. One is sure to begin with *Linnæus.* We have got the first volume of the new *Systema,* with very considerable additions. . . . . I doubt you will conceive that Linnæus is gone mad, if I tell you his opinion concerning *Funguses.* In a letter to Mr. Collinson, he thus expresses himself about them :—" Quis potuerat a priori dicere, Fungos esse *Animalia,* et eorum ova excludi in aquis, et more piscium ludere, dein transire in Fungos ? Mihi semper occurrit istud Plinii,—' mihi contuenti sese persuasit rerum natura, nil incredibile existimate de ea.' Delectatus fui hoc autumno videre istos vermes e quibus Fungi prodeunt, et eorum stupendam metamorphosin ex agilissimis vermibus in immobiles herbaceos Funges."—I shall soon begin to be in pain, lest our poor kingdom of *Vegetables* should be crushed into atoms, by the Animals on the one hand, and the Fossils on the other ! What Linnæus means, I do not at present understand ; but the very dreams of so great a genius merit our attention.

The system-madness is epidemical. Gleditsch of Berlin, and Crantz of Vienna, have both come out with natural ones, but I have not seen them. Adanson, in his new volume of "*Familles des Plantes*," has given us a sketch of about sixty new systems; altogether, in my opinion, not worth one farthing : he has, however, drawn up an account of former systems, which I think a good one ; if one can forgive his abuse of Linnæus, and his exaltation of M. Tournefort, which indeed is done in the true French spirit.

Tho. Martyn.

From Dr. Pulteney to Professor Martyn.

Blandford, 20 Jan., 1768.

Dear Sir,

I thank you particularly for the extract you gave me of Linnæus's letter to Mr. Collinson ; which was not only acceptable from the novelty of the opinion, but also as it reached me so long before I got the new edition of the *Systema*, where I see this opinion confirmed. It is, as yet, too strange to demand one's assent ; and is, I own, what I have yet no proper ideas of. The most obvious circumstance of a general nature that would lead one to assent to this opinion, is, as far as I can judge, taken from the natural or spontaneous, and chemical analysis of Fungi ; which, indeed, always shows more of an animal than vegetable decomposition. . . . .

R. Pulteney.

Among the literary persons whose intimacy Professor Martyn cultivated at Cambridge, was the late Thomas Twining, his fellow-collegian, the well-known translator[a] of Aristotle. In a letter[b] written long after Mr. Twining's death, and in the decline of his own life, the Professor thus adverts to the genius of his departed friend. " Thomas Twining was a man of the truest, best-natured humour, that I ever met with : indeed, he possessed it in so high a degree, that he appears to be the only one of my acquaintance who had this rare quality, said to be almost exclusively indigenous of Britain. Certainly the French, who have *beaucoup d'esprit,* have none of it. He had a most exquisite taste, both classical and musical : he was a profound Greek ; and to a thorough knowledge of the theory of music, he added a most chaste and correct execution on the violin. His ear was so nice and accurate, that he had acquired such precision in the French accent as to deceive a native : and yet he had never been out of England, nor had he conversed much with foreigners. I cherish his memory, both as he had many admirable qualities, and as he has afforded me many hours of rational pleasure and innocent cheerfulness." In the letter from which

---

[a] Thomas Twining was born in 1734. He was entered of Sidney College, and took his degree of M. A., in 1763. He was Rector of White Notley, in Essex ; and of St. Mary's, Colchester. He died in 1804.

[b] Addressed to Miss L. M. Hawkins, March 28, 1814.

this passage is quoted, Professor Martyn has preserved a little illustration of the "innocent cheerfulness" of his friend. Mr. Twining had no taste for the Botanical studies, which at this time began to revive in the University by means of the Professor's Lectures ; and still less for the anti-classical barbarisms, and the strange mixtures of Greek and Latin terms, which were freely adopted into the language of Botanists. In ridicule of what he esteemed a mongrel phraseology, he wrote the following extemporaneous jeu d'esprit,—in which he gravely, but very divertingly describes the College *Apple-Roaster*, after the manner of Linnæus :—

" MELOPTEUM stanneum, monopum, versatile : *seu* Ap-plé-ro-aster."

" *Corpore* versatili, circumactili ; fossulis stanneis, melo-decis, parallelis, intus distincto : *Cacumine* fastigiato, suggrundiato : *Latere* uno pendulo, circcumjectili, ambidextro, tinnitu crebro gaudenti ; alteris fixis, bullatis.

" Habitat in Conclavi Combinatorio seu Combibatorio Collegii Sidney-Sussex in Academia Cantabrigiensi."

" Now the misfortune is," adds Professor Martyn, " that, in order fully to relish the above description, it is necessary, not only to be acquainted with Latin, but to have seen the machine described, and to have some knowledge of Linnæus's manner. The *wycked Wyghte*, who was the author, was wont frequently to

lounge in my College rooms, and read Linnæus's works,
(which were then unknown in the University,) and
question me much about them."—It is much to be
regretted, that no Memoir of this distinguished
scholar has yet appeared. His brother, the late
Mr. Richard Twining, was once engaged in preparing
for publication the Remains and Correspondence of
his literary relative; but the design was interrupted
by his death, and has not been prosecuted by others.

His next publication was—" *A Sermon preached
before the Governors of Addenbrooke's Hospital, on
Thursday, June* 30, 1768, *in Great St. Mary's
Church, Cambridge :* by THOMAS MARTYN, B.D., *Fellow of Sidney College, and Professor of Botany in the
University of Cambridge.* 1768." 4to., pp. 10. The
text is, Isaiah liii., 4.

His learned father having died in 1768, he published,
two years afterwards, a selection of notes from his
unfinished MSS. on Virgil, under the title " Dissertations and Critical Remarks upon the Æneids of Virgil, &c. : by the late JOHN MARTYN, F. R. S., 12mo.,
pp. 227. 1770."—To this little volume he prefixed
" *Some Account of the late* JOHN MARTYN, F. R. S., *and
his Writings.* 1770." 12mo., pp. lxiii. and 227. This
is the Memoir which has been reprinted, with additions, in the first part of the present volume.

The same year produced another work, without his
name ; " *A Chronological series of Engravers, from the
invention of the art to the beginning of the present Century.* Cambridge, 1770."   12mo., pp. xii. and 128.

The Preface contains a Dissertation on the original
inventors of the art.   There are three plates of Engravers' marks.

An additional burden now devolved on the Botanical Professor, occasioned by the departure of Mr.
Charles Miller, the excellent Curator of the Cambridge garden, for the East Indies.   He was sent to
the Island of Sumatra, in 1770, to prosecute a favourite
speculation of Mr. Sullivan,—the finding and cul-
tivating nutmegs, or any of the spices, or vegetable
productions of the East, which might prove advantageous objects of commerce.   No deserving person
offering to succeed him, and the finances of the
garden being very low, Professor Martyn undertook
the office of Curator gratuitously, and performed its
duties several years, till the improvement of the funds
allowed the Trustees to appoint a salaried successor.

While thus engaged in the various duties of a College
Tutor, the Botanical Professor, the Walkerian Reader,
and the Curator of the Garden, he nevertheless found
time to send out a descriptive catalogue of the plants
under his care, with the title, " THOMÆ MARTYN, S.T.B.
*Coll. Sidn. Soc., Prof. Botan., Prael. Walk., et Hort.
Curat., Catalogus Horti Botanici Cantabrigiensis.* 1771."
8vo., pp. xi. and 193.   A portrait of Dr. Walker,
Founder of the Garden, forms the frontispiece: an
outline of the Professor's Botanical Lectures (pp. xi.)
is prefixed to the catalogue.   To this was added, the
next year, " THOMÆ MARTYN, *Prof. Botan. Mantissa*

Plantarum Horti Botanici Cantabrigiensis. 1772." 8vo.
pp. 31. A plan of the Garden is prefixed. These
Catalogues, which laid the foundation of the HORTUS
CANTABRIGIENSIS of the late Mr. Donn, (now rendered
so complete by Mr. Lindley,) were intended merely
for the use of his pupils. He was fully aware of their
deficiencies; but a manual of the Garden being
wanted for the immediate use of the students, the
Professor was under the necessity of compiling this
sketch "amidst a thousand avocations"—to use his own
language,—rather than leave their wants altogether
unsupplied.

Hitherto Professor Martyn had performed the duties
of his Professorship, of Dr. Walker's Reader, and of
the Curator of the Garden, without any emolument,
except the trifling return made by fees from the few
pupils who attended his Lectures. The Duke of Grafton, Chancellor of the University, being then Prime
Minister, was too sensible of his merits to allow his
valuable services to be overlooked. On the 2d of July,
1771, this distinguished nobleman signified to Mr.
Martyn, by the Bishop of Peterborough, then Master
of Trinity College, that he was disposed to procure an
endowment for the Professorship of Botany from the
Crown. Accordingly, he transmitted to his Grace,
at Euston, by Richard Croftes, Esq., M. P. for the
University, a certificate of his services, signed by
thirteen Heads of Colleges; accompanied by a
letter, of which the following is an extract :—

" The Professorship of Botany is now the *only* one that remains unendowed. If your Grace will please to honour me so far as to lay my case before His Majesty, with my humble prayer, for such encouragement as to his wisdom shall seem fit, I will continue my care of the Botanic Garden, and read annually, a complete Course of Lectures in Botany; adding every improvement to the science that is in my power, so as best to answer the purpose for which the royal munificence is requested." Unfortunately for Professor Martyn's interests, a change in the Ministry took place just at this time, which threw an obstacle in the way of this well-deserved endowment, and retarded it twenty-two years longer!

Writing to his friend, Dr. Pulteney, in February, 1772, he adverts with enthusiasm, to the delight with which he had recently " spent a morning with Mr. Banks and Dr. Solander, to turn over 3000 specimens of plants, 1000 of them new species, and coloured drawings of 700, all elegantly and accurately done upon the spot." " These gentlemen," he adds, " expect in less than a month, to set out for the Southern world; with three ships most royally equipped, and four draftsmen,—one for views and figures, the celebrated Zoffani—and three for Natural History."— " I envy you," replied Dr. Pulteney, (May 18th, 1772,) " the pleasure of your morning, spent in looking over the treasures of the New World, with our two philosophical circumnavigators. If these gentlemen should make an ample publication of their discoveries in natural history, how rich will they make us!—Shall we ever meet in this world or not? I know nothing that would give me equal pleasure with a journey to Cambridge. I have sometimes pleased myself with the thoughts of meeting you, by accident, in London." (See p. 178.)

Another work now made its appearance, in which Professor Martyn associated his labors with those of Mr. (afterwards Dr.) Lettice. " *The Antiquities of Herculaneum, translated from the Italian, by* THOMAS MARTYN, *and* JOHN LETTICE, *Bachelors of Divinity, and Fellows of Sidney College, Cambridge.* Vol. I. Part I. *Containing the Pictures. London,* 1773." 4to. pp. lxxiii. and 236, with 50 plates, 3 vignettes, and a map of the country about the bay of Naples. There is, also, a Catalogue of Pictures, Sculptures, &c., in the Museum at Portici; a Preface, and an Appendix, by the Translators. Some of the plates are beautifully executed. This work did not proceed any further, the sale being very limited. The Court of Naples made a formal remonstrance, by their Ambassador, to the English Court, against the publication of a work 'designed exclusively for *presentation* to Sovereigns, Princes, Noblemen, and Ambassadors!' Notwithstanding these lofty pretensions, and the fear expressed of a debasement of a Royal Italian Work, by its vulgar exposure in the literary market of England,—the Translators had themselves *purchased* the original for fifty pounds!

An important change now took place in Professor

Martyn's life,—occasioned by an event which vacated
his Fellowship, and withdrew him permanently from
the University except during the period of his Lectures.
On December 9th, 1773, he was united in marriage,
at All Saints' Church, Cambridge, to Miss Martha
Elliston*, sister to the Rev. Dr. William Elliston,
Master of Sidney College.   By this lady he had only
one child, John King Martyn ; who went to Sidney
College, and became a Fellow, and the Mathematical
Lecturer, there ;—but who soon relinquished his University prospects, for the more useful duties of the
ministry, and vacated his Fellowship by an early
marriage.   Professor Martyn's new connexion occasioned his retirement to the Village of Triplow, nine
miles from Cambridge, and contiguous to Foxton ;
with the sequestration of which he had been presented, by Dr. Keene, Bishop of Ely, the preceding
January.   Having entered on a married life with small
finances, he undertook the charge of two pupils, Mr.
Ingle, and Mr. Hartopp.   Providence interposed, at
this period, remarkably in his favour.   In College he
had recently cultivated a friendship with Mr. Warren,
(afterwards Sir John Borlase Warren,) which had
been promoted by his worthy acquaintance, the Rev.
Mr. Chappell, Mr. Warren's step-father.   On new-
year's-day, 1774, Mr. Warren, with the consent of
his guardians, Earl Ferrers, and Mr. Chappell,—he

* This lady survived the Professor, and lived to the very advanced age of 87 ; she died, August 27, 1829.

being a minor—presented Professor Martyn with the Rectory of Ludgershall, in Buckinghamshire; and the parties gave a further proof of their esteem and confidence, by engaging him to add Mr. Arnold Warren, the brother of his patron, to the number of his pupils. He remained at Triplow, till the autumn of 1776.

From Professor Martyn to Mr. C. Miller.

[Cambridge, Feb. 5, 1773.]

Dear Sir,

" Le Poivre says, that, in the Isle of Sumatra, and perhaps in the other Malayan Islands, there are mines of fine gold, called, in the language of the country, *Ophirs.* (See the extracts from the ' Travels of a philosopher, by M. le Poivre,' in the Annual Register, 1769.)—Query :—Whether upon inquiry in the country this would be found true ?—and, if so, Whether the word *Ophir* is a Maylayan word, and what is its origin, etymology, and import ?

" From *Ophir* (which some of the inquirers have placed in the Malayan Islands,—see Calmet ad verb. OPHIR,—) were brought by the ships of Hiram and Solomon, not only gold, spices, precious stones, and such productions as are found in the Malayan Islands at this day, but also the wood of a tree called *Algum* or *Almug.* See 2 Chron. ix. 10, 11, and 1 Kings, x. 11, 12.—This tree, whatever it was, grew upon Mount Lebanon, and was usually had from thence ;

and, from being mentioned along with the common
Cedar and *Fir*, (2 Chron. ii. 8.) is believed to be that
peculiar evergreen, now found growing in its natural
state no where else but upon Mount Lebanon, and
therefore peculiarly called the *Cedar of Lebanon;*
but which in no respect resembles the common berry-
bearing *Cedar*, which is usually classed among the
Junipers :—for the *Cedar of Lebanon, (Pinus Cedrus)*
produces cones, and is classed by Linnæus among the
Larch tribe.   This cone-bearing tree does not grow
in great plenty upon Lebanon itself, (see Miller's
Dictionary ;) and it was, in the days of Solomon, of
inferior quality to what was brought from *Ophir;*
(see 2 Chron. ix. 11., and 1 Kings, x. 12.) supposing
it to *be* the *Algum* tree above-mentioned : a word*
which the best Hebrew Philologists allow not to have

---

* [אלגום]. The Septuagint Version renders this word, in 1 Kings,
x. 12, Ξυλα πελεκητα, *Trees cut by the axe ;* and in 2 Chron. ii. 8.,
2 Chron. ix. 11., Ξυλα πευκινα, *Pine* trees : but different MSS. read,
Απελεκητα, *uncut by the axe ;* Αρκευθινα, probably *Junipers ;* and
Γουγειμ, or Αγουγειμ, *Gougeim*, or *Agougeim.*   Josephus renders it,
Ξυλα πευκινα, *Pine* trees, but says it was different from the πευκη, or
Pine, known in his time ; the wood being ' like that of the fig-
tree, but more white and shining.'   St. Jerome interprets it, *Ligna
Thyïana,* by which he seems to mean a species of *Thuja,* or
Arbor-Vitæ.   It thus appears how little is to be gathered from
the interpreters.—The kind of investigation hinted at by Professor
Martyn, if diligently pursued by persons competent to the task,
would be more satisfactory than the mere guesses of commentators.]

L

been of Hebrew original, but probably imported with
the tree. (See Lexic. Hebr. Reckenbergeri, Jenæ,
1748, 8vo.)—Query: Whether the *cone-bearing Cedar*
is found* in the mountains of Malacca, Sumatra, &c. ?
and, Whether it is there called by any name like
Algum or Almug ?"

THO. MARTYN.

From Mr. C. Miller to Professor Martyn.

Fort Marlbro' [Sumatra,] Jan. 7, 1774.

Dear Sir,

Agreeably to your desire, I have made inquiry
among the most learned of the Malays, and also
among those of the several different nations on this
Island, concerning *Mount Ophir*, and the *Almug Trees*.
No such words as *Ophir*, or *Almug*, are known to any
of them.   M. Poivre is, therefore, greatly mistaken
in asserting, that the gold mines on Sumatra, are
called *Ophirs*: they are called by the Malays, Tom-
bong, and are known by no other appellation.   He
has probably been led into this mistake by seeing
a mountain, on the northern part of this coast, called

* [Mr. Miller takes no notice of this point, in his reply: but he
had remarked, in a previous letter, (March 8, 1771,)—"the
country is very difficult to penetrate, on account of the *Rattan*,
(*Calamus Rotang*,) which is very full of thorns, and grows across
from tree to tree." He observed, in another letter, (May 9, 1773,)
" I have met with very few trees in these woods but what were
entirely new and undescribed."]

Mount Ophir in the charts, and by being told that it yields gold and Cassia. This, however, is not the country name for that mountain; but one wantonly given to it, perhaps by some of the early navigators on this coast. I think it is hardly probable, that any vessel made such a long voyage in those early times; and I should much sooner look for the *Ophir* of Solomon, on Ceylon. . . . .

CHARLES MILLER.

In the autumn of 1775, Professor Martyn made a journey to the west of England, and visited the Isle of Lundy, in the mouth of the British Channel, the property of a young gentleman, with the management of whose affairs he was entrusted. He had much wished to visit his scientific friend, Dr. Pulteney, *whom he had never yet seen;* but the journey being exclusively one of business, he was unable to digress into Dorsetshire:—thus these two eminent Botanists, seemed fated to have no more than an epistolary intercourse! This year he published, " *Elements of Natural History. By* THOMAS MARTYN, *B. D., Professor of Botany in the University of Cambridge. Cambridge, 1775.*" 8vo. pp. 70. This was dedicated to Thomas Pennant Esq. It is the *first* part only, containing the " *Mammalia.*" The author was furnished by Mr. Pennant with his MSS. notes on Birds; but the work was not carried on.

From Dr. Pulteney to Professor Martyn.

Blandford, Oct. 3, 1775.

Dear Sir,

. . . . I have no Botanical news to send you.
My distance from London, and the little correspondence I have with naturalists, keeps me in a state of
great poverty in these things. The following, as it
occurred to myself, is the only anecdote that I can
think may be acceptable, *if* it should prove new.
Mr. Lightfoot, of Uxbridge, did me the favour, in
August, 1774, of spending three or four days with
me, in his way into Cornwall, on a *simpling* expedition; in company with the Rev. Sir Henry Parker,
Bart., of Trinity College, Oxford; a great lover of
Botany. I had the pleasure of showing them, for
the first time, in their native places of growth, the
Pinguicula villossa[a], *Gentiana filiformis*[b], and *Œnan-
the Pimpinelloïdes*[c]. I was almost tempted to execrate
my profession, that would not suffer me to accompany
them to Cornwall. They had a very successful journey, after they left me. . . . Mr. L. was fortunate
enough to add two new plants to the British Flora.
In Portland, he found the *Polycarpon tetraphyllum*[d];

[a] [*P. lusitanica*, Pale Butterwort, of Eng. Flora.]
[b] [*Exacum filiforme*, Least Gentianella, of English Botany.
t. 235, figured from a plant sent to Mr. Sowerby by Dr. Pulteney.]
[c] [Parsley Water-Dropwort.]
[d] [Four-leaved All-seed.]

and, near Axminster, the *Lobelia urens*[a]. Among many other fine plants, they saw great plenty of the *Erica multiflora*[b], which occupies five miles of the environs of the Lizard Point.—One considerable object of his journey was finding *Smyrnium tenuifolium*[c], Raii Syn. 209, of which he had seen the specimen, in Buddle. He did *not* find any plant that answered to it. . . .

> RICHARD PULTENEY.

At Triplow, Professor Martyn found himself happily settled in a quiet retirement, very congenial to his mind; and yet he was not too far from his University friends, to enjoy their occasional society. But the further kindness of his patron, Sir J. B. Warren, tempted him to remove to Little Marlow, in Buckinghamshire, as the Incumbent of that Vicarage, in the autumn of 1776. This preferment rendered it necessary for him to resign the sequestration of Foxton. At Little Marlow, (where his residence was very much interrupted,) his engagements were so many, that he found scarcely any time for his favourite science. " Nothing new,"—he writes to Dr. Pulteney, (Aug. 6, 1777,)—" has occurred to me here; except that *Daphne Mezereum* grows wild commonly in our woods."

[a] [Acrid Lobelia.]   [b] [*E. vagans*, Cornish Heath.]
[c] [*Ligusticum Cornubiense*, Cornish Lovage.—Though Lightfoot failed in his search for this rare plant, it has since been found, near Bodmin, by Mr. E. Foster.]

In the autumn of 1777, he removed, with his ward and pupil, Mr. Edward Hartopp, to the family seat, Little Dalby Hall, in Leicestershire; where he spent a part of the winter. The state of Mr. Hartopp's health rendering it advisable for him to change the climate, his tutor was invited to accompany him for two years to the Continent. Accordingly, on the 1st of August, 1778, Professor Martyn left England, accompanied by Mrs. Martyn, his infant son, and his pupil. Travelling by Calais to Paris, (where they made a short stay,) they proceeded through Lyons to Geneva. They resided, during the month of September, at Vandœuvre, a beautiful little villa, between three and four miles east of the city, and a mile south of the lake. From this place, Professor Martyn addressed the following letter to his friend, Mr. Ashby.[a]

Vandœuvre, Oct. 1, 1778.

Dear Sir,

I have taken the opportunity of an English gentleman's returning home, to send you a packet, on the subject of Cæsar's operations. . . . By means of Cæsar and the map, and a conversation with my friend, Mr. Lettice[b], who has resided at Geneva some time, and Mr. Kerr, Sir John St. Aubyn's tutor, I

[a] See above, p. 110.

[b] [Now Dr. Lettice, Rector of Peasmarsh, in Sussex. See above, p. 142.]

had entirely satisfied myself that you were in the right, with regard to the place of Cæsar's works[a], when I met with Dr. Butini's dissertation, which I have translated and sent you.

The stupidity of commentators, and the inaccuracy of travellers is astonishing. I think myself so much *au fait* with regard to this matter, that if you and I could traverse the Alps together, at leisure, I have no doubt but we should follow Hannibal's track without any great difficulty; and if I return hither next spring, I will try what can be done on *that* subject, if you will favor me with your hints and instructions.

On the subject of Cæsar's operations, I have only to add, that the Rhone is fordable[b], as I am informed, in several places between Geneva and the

---

[a] [*Cæsar's wall.* See Cæs. de Bello Gall. Lib. I. 6—8. Spon, in his *Histoire de Geneve*, thinks it was between *Gingin*, (near *Nyon*,) and the lake of Geneva. But Dr. Butini, in his *Dissertation sur le lieu, ou passoient les lignes, que Jules Cæsar fit faire près de Geneve*, (appended to the second Vol. of Spon, 1730,) maintains that the wall, or vallum, extended from *Geneva* to *la Cluse*, on the *Rhone*, and the mountain now called *le Wache*, but which then, as is supposed, went under the name of the *Jura*.]

[b] [He adds, in a letter from Cambridge, August 30, 1781,—" I apprehend the Rhone to be fordable in many places between *Geneva* and *Fort la Cluse*; but then the banks are so very high and steep, that it would be impossible to effect the passage in face of an enemy." The subject seems to have been fully discussed between the Professor and his correspondent, in the period 1778—1781 ; but the letters are lost.]

Vuache; that considering Cæsar's alacrity, there appears nothing incredible to me, in his raising a parapet 18,500 paces in length, in a fortnight's time, with the assistance of the country people; and that the march of 100,000 men, with women and children, much baggage, and three months' provision, would give Cæsar time to assemble fresh forces, and meet the Helvetii as they were crossing the Arar*. The Arve, is as wide as the Rhone after junction. The reason is, that the former flows through a cultivated country; whereas the latter is confined in a rocky channel, and is deeper.

Our parlour windows have the mountains of Savoy in full front, with the glaciers towering over them, exhibiting a most august view at sunset, in a clear evening. When we have rain, as soon as the sky clears up, we perceive snow on the tops of the second ridge of mountains from us; the third ridge covered with snow, all which has hitherto disappeared in a few days; and beyond all, the vast mounds of eternal ice. My abode here has given me a most contemptible idea of the clouds, which we continually see traversing the sides of the mountains, leaving the upper parts visible.

I heartily wished for you in the narrow defile from from Nantua to la Clûse; you would have been, at the same time, charmed with the road and country,

* [The Soane.]

and convinced of the difficulty the Swiss must have
had to get along such a narrow way. . . .

THO. MARTYN.

During October, the party made an excursion to
Lausanne, Vevay, and the salt mines at Bex; and,
on the 18th of November, established themselves in
Geneva for the winter.

" As long as the winter lasted,"—writes the Professor, in his MS. Journal, (pp. 180, 181,)—" we remained patiently within the walls of Geneva, partak-
ing of the excellent society which that city affords:
but no sooner did spring begin to make nature gay,
than we longed ardently* to put in execution a plan,

---

* [Professor Martyn's MS. Journal, contains abundant proof of
the ardour with which he pursued his way, amidst scenery, (to use
his own words)—

"Where darksome pines arrest the wand'ring eye,—
    Flowers of all hues enamel the gay ground ;—
The *Alpine Rose,* displays her sanguine dye,
    And Aromatics fling their fragrance round."

The plant to which he here alludes, is, the *Rhododendron ferrugineum,* commonly called the *Mountain Rose ;* one of the most
beautiful and abundant of shrubs on the Alps ;—the *Rhododendron hirsutum,* (a species nearly allied to the former,) is found
less commonly, and is not so deep in its " *sanguine dye.*"—" In
the high mountains of Switzerland," observes the Professor, (in
Miller's Dictionary,) " the *Rhododendron ferrugineum,* and *R.
hirsutum,* terminate ligneous vegetation, as we ascend, and furnish
the shepherds with their only fuel. I found the *R. hirsutum* in
great abundance, in August, 1779, on Mount Sheidegg, in crossing
to Grindelwalde."]

which we had meditated, of a tour through Switzerland. Such were our expectations of pleasure from it, that we could hardly persuade ourselves to wait for the proper season for such an excursion; a great part of which cannot conveniently be made but during the summer. In the mean time, I amused myself very well, during the months of March and April, with several Botanical excursions in the environs; particularly to the wood of La Batie, Mont Saleve, and along the side of the Rhone to Chatelaine. On the 14th of March, in a most delightful walk to Mont Saleve, by Bossay and Veri, over the mountain, by Montroux and Mornex, and back by the Pont d' Estrembieres and Chêne—I found the spring plants in flower; as, *Anemone Hepatica*[a], *Scilla bifolia*[b], *Viola martia*[c] with white flowers as well as purple, *Viola hirta*[d], *Potentilla verna*[e], and *Fragaria sterilis*[f]. The day following, at the wood of La Batie, on the banks of the Arve, I saw, for the first time, the beautiful flowers of *Erythronium Dens-Canis*[g]. I found, also, *Anemone nemorosa*[h], and *Pulmonaria officinalis*[i]. On the 23d of the same month, I saw *Daphne Mezereon*[k],

---

[a] [Common Hepatica.]   [b] [Two-leaved Squill.]

[c] [*Viola odorata*, (Sweet-scented Violet,) of Linnæus.]

[d] [Hairy Violet.]   [e] [Spring Cinquefoil.]

[f] [*Potentilla Fragariastrum*, (Strawberry-leaved Cinquefoil) of the Eng. Flora.]

[g] [Dog's Teeth Violet.]   [h] [Wood Anemone.]

[i] [Common Lungwort; a rare plant in England.]

[k] [Mezereum, or Spurge-Olive.]

in full bloom.—But I have no design to write the natural history of this country, and have set down these plants chiefly with a view to compare the time of their flowering here and in England."

On the 19th of May, 1779, the party left Geneva, and proceeded through the following places;—Lausanne, Vevay,* Aigle, Bern, Soleure, Bâsle, Schaff-hausen, Constance, Zuric, Lucerne, Bern, (for the second time,) Bienne, Neufchatel, Bern, (for the third time,) Thun, Brientz, Meiringen, Grindelwald, over the Sheidek, Lauterbrunnen, Bern, (for the fourth time,) Fribourg, Martigni, Chamouni by the Tête Noir, Martigni by the Col de Balme, Vevay, and Neufchatel, (the second time);—they returned to Geneva on the 19th of September.

On the 5th of October, (1779,) they left Geneva for a tour in Italy; and crossed Mont Cenis, to Turin, whence they proceeded to Genoa, Pisa, Leghorn, Florence, Rome, Naples; and returned to Rome, on the 2nd of March, 1780. Leaving that city, on the 28th of May, they proceeded through Bologna,

---

* " At Vevay, and Aigle, I observed that many of the *Alpine* plants occur in the beds of the torrents; the seeds being, doubtless, brought down along with the water that pours from the mountains. —The fruits were now coming into season : cherries were sold at Geneva, (May 18th) the day before we left it; we had strawberries at Lausanne on the 20th of May; and on the 24th, the grapes of last season, and the strawberries of this, met in our dessert." (Martyn's MS. Journal, Vol. I., pp. 192, 193.)

Ferrara, Venice, Verona, Trent, and through the Tyrol to Augsburg, Ulm, Stutgard, Manheim; down the Rhine to Cologne; thence to Spa, Liege, Brussels, Ghent, and Bruges; and, embarking at Ostend for Margate, they reached England on the 2d of September. After a short visit to his Vicarage, at Little Marlow, the Professor went with his family to Little Dalby-Hall, the seat of Mr. Hartopp, to arrange his affairs, and to prepare that place for his pupil's future residence.

Professor Martyn kept a full Journal of this tour, which he afterwards transcribed in a very fair hand, in two 4to. volumes, comprising 1040 pages, and entitled, " *Travels through Switzerland and Italy, &c., during the years* 1778, 1779, *and* 1780." This Journal exhibits proofs of the diligence and accuracy of his observations as a traveller; and it is written in an easy, interesting style. It seems to have been drawn up entirely for the private use of himself and his friends; but, being urged to publish some account of his travels, he at length in part yielded to repeated solicitations, and sent forth a small volume containing a few *selections* from his MS. Journal, consisting chiefly of details relating to painting, sculpture, architecture, the best routes, &c. &c. This volume will be noticed below.* It is to be regretted that his MS. contains scarcely any remarks connected with Bo-

---

* See under the year 1787.

tany; for which omission his own apology may be seen in some of the following[b] letters. He was not, however, forgetful of his favourite science; he brought home a rich *Hortus Siccus*, and many beautiful drawings of plants.

From Professor Martyn to Dr. Pulteney.

London, April 6, 1781.

Dear Sir,

   . . . You will suppose that the Swiss mountains were highly interesting to me. Indeed they were a most rich repast, and I was not idle among them; for I made a copious Hortus Siccus; collected a great variety of fossils; and sent over a large case of rare Alpine plants to the Botanic Garden, at Cambridge. In Italy my pursuits were chiefly of another kind; the beautiful remains of antiquity employing a great share of my time and attention. The heat of the climate, and the weakness of my constitution, forbad me to make many Botanical excursions; my residence was chiefly in great cities; and I was not able to accomplish a journey into Calabria and Sicily: so that I made but inconsiderable additions there to my Hortus Siccus. I collected fossils, however, in abundance, especially those which are volcanic. I

[b] See the letters dated April 6, 1781, October 26, 1781, and January 16, 1783—*infra.*

have already received, besides my Swiss cases, one from Naples, and four from Rome; and I am in daily expectation of three from Venice. These will greatly improve my Museum, and furnish abundance of new materials for my Lectures, which will begin on the 26th instant. I should delight much in having an opportunity of displaying to you my treasures; and beg that, if any of the Swiss plants are interesting to you, you would turn over your Haller, and command any duplicates among my specimens. . .

THO. MARTYN.

From Dr. Pulteney to Professor Martyn.

Blandford, 26th April, 1781.

Dear Sir,

I thank you for announcing to me your return to England, and most sincerely congratulate you on having so happily effected an excursion, in itself desirable in the highest degree, to one of a cultivated taste for such objects, as are more immediately your pursuit. The practice of *Physic* confines its votaries to a spot, and admits of so little recreation at a distance, that, unless every gratification comes *home* to us, we are desolate. No man can have had stronger wishes than mine have been to enjoy the pleasure of exotic rambles; but fate has denied me the completion of them, till necessity has done that which philosophy ought to have perfected before :—

thus, therefore, I submit. . . . . With what delight I should have accompanied you to the places that have known the footsteps of the illustrious HALLER, let your own enthusiasm tell!

RICHARD PULTENEY.

From Professor Martyn to the Rev. G. Ashby.

Cambridge, Aug. 30, 1781.

Dear Sir,

　　I have made some progress in the translation of M. de Saussure; but I am so great a vagrant, that little can be done either in the writing or reading way.

　　When I am settled, it is hard if I cannot translate a quarto, that I have calculated may be gone through in six weeks, at six hours per day.　After all, none of the million, that admire Moore, Coxe, Wraxall, and such pretty writers, will read De Saussure; physics and natural history not being quite so universal as namby-pamby, and anecdote.

　　I remember to have seen Thoughts on Rowley in the Magazine, to have known them for yours, and to have been satisfied with what you said.—I have long intended to look over Stevens's Shakspeare, with a view of knowing what was done by the editors, and whether I had any observations by me which they have not—mine principally relate to the natural objects he occasionally mentions, and his allusions to natural history—but alas, I have never yet found time.

THO. MARTYN.

From Professor Martyn to Dr. Pulteney.

Easton Lodge^a, Essex, Oct. 26, 1781.
Dear Sir,

I was much flattered by a present of your interesting publication^b . . . It afforded me particular pleasure to find you giving an account of what Linnæus had done for the improvement of natural history; because, since his death, some *minute* naturalists have been plucking at him, and endeavouring to depreciate his genius and labours : even men who cannot write English have abused him for not having a Ciceronian^c style ! . . . .

My Journal will not furnish any thing complete enough for the eye of the public.   I was, indeed, naturally led to make some observations upon public institutions, and the state of natural history, wherever I resided ; and I could easily say a great deal upon how much is done with very small means in Switzerland, and how little with ample means in Italy : but I travelled

---

^a [The Seat of Viscount Maynard.]

^b [" A General View of the Writings of Linnæus. 1781." This valuable work was republished by Dr. Maton, in 1805, with the Life of Dr. Pulteney prefixed.]

^c [Rousseau, however, did not scruple to say, that if *Cicero* had described plants, he would have written *like Linnæus !*]

with a family, and a pupil, and, therefore, had not
leisure or means to pursue a plan of my own; being
frequently subservient to the caprices of others, and
obliged to submit my plans to theirs. . . . .

I have not been settled any where since my Course
of Lectures finished in June last, but have been visiting in different parts of the country. We now think
of a journey to Bath for about six months. . . . .

Tho. Martyn.

About this time he seems to have paid a good
deal of attention to Mineralogy. Among Sir Joseph
Banks's MSS., in the British Museum, there is a
letter addressed to him by Mr. E. M. Da Costa,
(dated July 1, 1782,) acknowledging the receipt of
a box of very rare fossils, from the chalk pit at
Cherry Hinton, near Cambridge.

He now furnished his pupils with a new Syllabus of
his Lectures, (which had been much more extended
in their plan than when he published the former Outline of his Course, in 1764,—) under the title,
" *Heads of a Course of Lectures in Natural History,*
read at the Botanic Garden;' by Thomas Martyn, B. D ,
Professor of Botany in the University of Cambridge.
Cambridge, 1782." 12mo., pp. 44. These Lectures
are arranged under the divisions of I. *Animals,*
classed after Linnæus, with the improvements of
Pennant in the Quadrupeds and Birds; II. *Vegeta-*
bles, classed according to Linnæus; III. *Fossils,* after

M

the method of Professor Wallerius, of Upsal, whose system had been published four years before.

From Professor Martyn to the Rev. G. Ashby.

[*Not dated.*] Cambridge, March, 1782 ?

Dear Sir,

Dr. Pallas's account is very satisfactory. I am glad to find any fact which so effectually overthrows M. Bailly's reveries; for I can call them no other. Poor little England! so we were vomited up the other day. But how is it that no marks of subterraneous fires have been detected on our island; except* in the farther parts of Scotland, perhaps in Wales, and probably in Derbyshire? Whereas in every part of it, we see marine sediments, and the effects of water, with the exuviæ of marine and other animals, mixed with fragments of vegetables, of no very recent date. But we are not contented now-a-days with throwing about Pelion, Ossa, and Olympus—no! we manage whole islands, nay, whole continents, with as much ease as a boy does his marbles; now blowing them all up in the air; and now drowning them in the ocean!—After all, my dear friend, we are mere children in these matters, notwithstanding all our vaunts. We possess a few facts; but as soon as we come to reason upon them we grow ridiculous.

* These *exceptions* ought to be considerably extended.

I suppose you know that Coxe is printing a volume upon the Grisons; and that Miller is indignant beyond measure, that the extracts from his letters have been printed.*

THO. MARTYN.

From Professor Martyn, to Dr. Pulteney.

Little Marlow, Jan. 16, 1783.

Dear Sir,

The pleasure it would have given me to meet with you at Bath, would have been very great indeed; but I begin to think we are destined *not* to meet!

I look upon the work you mention, viz., the modelling Linnæus's Botanical works over again, as the great desideratum of Botany. Dr. Solander was certainly the man who could have executed it best; and is, on that account, as well as many others, much to be lamented. Such a work would be highly worthy the attention of the President of the Royal Society, [Sir Joseph Banks;] but I am not acquainted with any person who is *equal* to it, and at the same time has *leisure.*

I find every day less inclination to publish any thing in consequence of my tour,—never having had such a thing in view, not having staid long enough in a place to make a sufficient quantity of accurate ob-

---

* [In the Philosophical Transactions. See above p. 114.]

servations, and having been encumbered with a family, a pupil, and abundance of cares concerning eating—drinking—lodging—travelling—and etiquette. Little, indeed, has been done in *our* way, by modern English travellers; but then Ferber, Saussure, De Luc, &c., have made some amends in the countries I passed through. I had translated half Saussure, [*Voyages dans les Alpes*;] and had thoughts of putting together what the three above-mentioned authors have given us, with some observations of my own : but I find no time to execute the plan I had conceived.

This year has been employed in revising, correcting, and adding to my Lectures. Wallerius's own edition of his Mineralogy in Latin being very difficult to obtain, I had thought of translating it into English. . . . . .

THO. MARTYN.

From Professor Martyn to Dr. Pulteney.

Little Marlow, July 30, 1783.

Dear Sir,

. . . I presume you know that Da Costa is about a work which he calls " *A Synopsis of British Fossils ;* " but he is a Mineralogist of the last age. . .

You are in correspondence with Mr. Relhan. His knowledge and diligence is such that I hope he will give us a " FLORA," which will do us credit. I exhort him to print somewhat in the form of Scopoli, and to give a neat description of each plant.

Will it not be remarkable if Cambridgeshire should have *four** Floras, before Oxfordshire has *one* ? Is it not, also, remarkable, that when Mr. Hasted designed to give a catalogue of the wild plants of Kent, he should be wholly unacquainted with *Hudson's Flora*, and even *Ray's Synopsis* ?—and that, to the last, he should be ignorant of old *Johnson's Itinerary* ? It is not necessary that every antiquary should be a naturalist; but, if he steps aside into *the flowery path*, he should at least procure a friend to direct his steps.

I am here till the winter; then, whether I shall be here or in London is yet uncertain ;—*wherever* I may be, depend upon it, Dear Sir, I am, with the most perfect respect and esteem,

Your faithful servant,

THO. MARTYN.

In the following year, (27th May, 1784,) he was elected a member of the " Society for the promotion of Philosophy and of General Literature," which had been recently established at Cambridge. It may not be uninteresting, to the Academical reader at least, to be presented with a short account of an Institution which, though formed by some of the most distinguished persons in the University, did not meet with

---

* [Ray's, in 1660 ; Professor John Martyn's, in 1727 ; Professor Thomas Martyn's, in 1763: and Relhan's in 1785.—Sibthorpe's *Flora Oxoniensis* was not published till 1794.]

the success which had been anticipated; it is due also
to the memory of the individuals who exerted themselves on this occasion, that their zeal in the cause of
science should not be unrecorded.   This Society was
projected in 1782; but the death of some persons
interested in the plan, and several accidents, occasioned the scheme to be postponed till February 18th,
1784, when the following gentlemen associated themselves under certain laws and regulations :—Professor
Milner, (afterwards President) of Queen's College,
Mr. (afterwards Archdeacon) Coxe, Professor Jowett,
Mr. (afterwards Professor) Carlyle, Mr. Atkinson, Dr.
Coulthurst, and Mr. (now Professor) Farish.   They
soon added to their number, Mr. Pearce, Mr. (afterwards Professor) Vince, Sir Busick (afterwards Professor) Harwood, Mr. Relhan, Mr. Jones, Mr. (afterwards Professor) Porson, Mr. I. F. F. Emperius,
Professor Martyn, Mr. Popple, Mr. Brundish, Mr.
(afterwards Professor) Tennant, Mr. (afterwards Professor) Wollaston, and Mr. Ainslie.   This little Society of learned men, not being adequately supported,
was dissolved* about the close of the year 1786.   It
had been resolved, by its distinguished members, that
none of them should be idle; but that each should
occasionally furnish original papers, of which a selection should be printed in a volume, to be entitled

* A similar attempt was made under more favourable circumstances, 35 years afterwards, in the establishment of " The Cambridge Philosophical Society," in the year 1819.

" Tracts, Philosophical and Literary, by a Society of gentlemen of the University of Cambridge." Of the Essays* which were thus distinguished, one was communicated by Professor Martyn, and is entitled, *" An enquiry into the nature and use of Pozzolana earth; its several substitutes, and the places where it may possibly be found, in Great Britain and Ireland. By* THOMAS MARTYN, *B. D., Professor of Botany in the University of Cambridge."* 4to. pp. 18. [Ordered to be printed, May 12, 1785.] This Paper forms pp. 35—52 of the proposed volume, *which was never published,* being incomplete. Professor Martyn's Paper is preserved in the Library of Queen's College, in the valuable collection of Tracts by the late Dean Milner, in the volume marked Gg. 1. 42. The Professor considers *Pozzolana* as an argillaceous earth, formed by the pulverization of various rocks in the crater of a volcano. Being mixed with certain proportions of lime and rough sand, it sets firmly under water. In the

---

* The other Essays, ordered to be printed, were, I. " *On the Summation of Infinite Series, by* S. VINCE," Dec. 30, 1784 ;— II. " *On Centripetal forces, by* Dr. WARING," Dec. 29, 1785 ;— III. " *On the precession of the Equinoxes, by* Mr. VINCE," Feb. 23, 1786 ;—IV. " *On the Ascent of an Air Balloon, by* Mr. JONES," April 6, 1786 ;—V. " *On the Limits of Equations, by* Mr. MILNER," May 11, 1786 ;—VI. " *On the Motion of the Moon's Apsids, by* Mr. MILNER," June 1, 1786.—Dr. Waring's Paper is preserved, in the same Volume of Tracts which contains Professor Martyn's Essay: it is not known whether any printed copy of the *other* Dissertations exists.

old tower of Toulon, he observes, (from M. Faüjas de St. Fond,) which is built in the sea, while the salt water, and the force of the waves, have decomposed the stone, the *Pozzolana* cement remains. It forms excellent floors, and incrustations for the roofs of houses.

Family circumstances (and particularly the education of his son, under Dr. Smith, at Westminster) now rendered it desirable for Professor Martyn to remove to the Metropolis. Accordingly, he quitted the Vicarage of Little Marlow, in November 1784; and before Christmas he entered upon a residence at Park Prospect, Queen Street, Westminster. Early in the ensuing year, (1785,) he purchased Charlotte Street Chapel, Pimlico, from Dr. Dodd; and he began his ministerial duties there on Easter Sunday. On the 4th of June he resigned the Rectory of Ludgershall; in which living his half brother Claudius became his immediate successor.

By far the most popular of all his works made its appearance this summer; viz., his *" Letters on the Elements of Botany, addressed to a Lady, by the cele- brated* J. J. ROUSSEAU ; *translated into English, with Notes, and twenty-four additional Letters, fully explain- ing the System of Linnæus: by* THOMAS MARTYN, B. D., *Professor of Botany in the University of Cam- bridge. London, 1785.*" 8vo., pp. xi. and 503. This interesting volume has been through eight editions. It was rendered much more useful, three years after the date of the first impression, by the publication of

" *Thirty-eight Plates, with explanations; intended to illustrate Linnæus's System of Vegetables, and particularly adapted to the Letters on the Elements of Botany :* by THOMAS MARTYN, B. D., F. R. S. & L. SS., *Professor of Botany in the University of Cambridge. London,* 1788." 8vo., pp. 72. These plates were well engraved by Nodder, and should always accompany the original work; they are also well adapted to the illustration of other Introductions to Botany. The Letters are written in a remarkably easy, perspicuous, and engaging style; and the particular plants described are admirably selected for the illustration of the Classes. Only the Introduction, and eight of the Letters, are from Rousseau; the remaining twenty-four, besides several notes to the preceding, are from the pen of Mr. Martyn. It is, perhaps, to be regretted, that the modest author did not make his volume *entirely* original; for it may be unhesitatingly said that Rousseau's part of the work (however elegant in style) is the *least* valuable portion of the volume. The name of the Swiss, however, generally usurps the place of that of his infinitely more learned Translator and Continuator; this work being, absurdly enough, usually called " *Rousseau's* Letters on Botany !" Mr. Martyn was employed exactly one year in the preparation of this work. In transmitting a copy of it to his friend, Dr. Pulteney, he modestly says—(June 22, 1785,) " I send it merely as a testimony of my respect and gratitude; for it cannot be of any use to *you*, nor even afford you any amusement: if you have any friend less informed than yourself, to whom it may be of any use, I shall be satisfied." Dr. Pulteney, however, was of a very different opinion. Dr. Darwin, also, did not hesitate to acknowledge that he had derived both " pleasure and improvement," from a volume which he considered as well calculated to " propagate the science[*]."

The old Lecture-room in the Botanical Garden having been pulled down, and the new apartment not being yet ready, the Professor omitted his usual Course this year; for which, indeed, he would have been otherwise unfitted, by the state of his health.

All the literary or scientific undertakings, in which Professor Martyn had been hitherto engaged, were upon a comparatively small scale, and were within the compass of ordinary diligence. But he now determined to employ his time and talents upon a work as great, perhaps, as was ever executed by any *one* individual; a work which gave him incessant occupation for twenty-two years, and which must be considered as the monument of his Botanical fame. This was a new edition of Philip Miller's " *Gardener's Dictionary;*" or, rather, an entirely new work, founded on Miller's as its basis, embodying all the discoveries of Linnæus in the vegetable kingdom, together with all information of importance for the Botanist, the Gardener, and the Husbandman, which existed at the

* Letter from Dr. Darwin to Professor Martyn, Derby, Feb. 6, 1786.

close of the 18th Century! Mr. Curtis had been engaged for this work by the booksellers; but, after having had it in hand for three years, relinquished it without having done any thing. Messrs. White and Rivington then engaged Professor Martyn to enter upon the Herculean labour in the winter of 1785. The remuneration agreed upon was 1000 guineas; a sum which some of the publishers "thought enormous," but which the author himself considered not very advantageous, as he expected to be occupied eleven years in the work: in point of fact, this sum was a very inadequate compensation for his labours, for twice that period elapsed before he completed his vast undertaking! The work was commenced Nov. 11, 1785; the first sheet was received from the press Dec. 29, 1792; the publication of the first part (containing forty sheets) took place May 30, 1795; and the whole was published Dec. 21, 1807, in four very large folio volumes, (—titled as two volumes, in two parts—) price fourteen guineas. We shall revert to our account of these volumes, and give a more full description of their contents, when we come to treat of that period of the Author's life in which he completed his great work. In the mean time, the progress of his labours will be occasionally noticed in extracts from his correspondence, in which he freely solicited the hints and criticisms of his scientific friends.

To the Gentleman's Magazine for 1785, (Vol. LV.,

Part II., p. 757,) he contributed a paper, with the signature P. B. C., containing some *Suggestions on the utility of publishing a Catalogue of plants, with the names accented, and observations on the disputed pronunciation of several names.* This had already been partially done by the Lichfield Society, in the Index of their translation of Linnæus's " Systema Vegetabilium ;" Professor Martyn himself prosecuted this plan in his " Gardener's and Botanist's Dictionary ;" and it is now generally adopted.

From Dr. Pulteney to Professor Martyn.

Blandford, Jan. 4, 1786.

Dear Sir,

" I am glad that Miller's book has fallen into your hands, and doubt not that you will do ample justice to the work of the good old man. The retrenchments that may now safely be made, will enable you to introduce all the new plants, and make the work, as far as the Botanist is concerned, *perfectly Linnæan.* It will, by this means, greatly tend to familiarize the system among all ranks of people; a circumstance much to be desired.

" If I do not greatly mistake, I recognize in the Gentleman's Magazine for October last, under the signature P. B. C., the Cambridge Professor. If I am right, suffer me to intreat that you will not let that matter sleep; but, if possible, take it into your own

hands, and take in the generical Synomyma of other
Authors : it cannot but be an acceptable thing to the
public as you will manage it." . . . .

RICHARD PULTENEY.

After having corresponded for twenty-six years
without a personal knowledge of each other, these
two Botanical friends at length met, accidentally, in
the shop of Mr. White, the bookseller, in Fleet
Street! (See p. 142).

From Professor Martyn to Dr. Pulteney.

Park Prospect, July 26, 1786.

Dear Sir,

I cannot neglect the opportunity of Mr. Hudson's journey into the West, to pay you the only
respect I am able to do, and to inquire after your
health, which I hope is re-established.   I am sorry to
find that Mr. Hudson's is no better.   The health of
those who know how to employ their time, so well as
he and you do, is a public concern.   A return of *his*,
would, I hope, produce us a third edition of the
" Flora Anglica ;" and I sincerely wish, for the advantage of our favourite science, that *you* could drop the
fatigue of business, and attend to it wholly. . . .

The " Dictionary" advances ; but I shall find it a
long and heavy business.   Besides the addition of all
the new Species, the Generic and Specific characters

must all be translated anew; and I am determined that the book shall contain in English *the marrow* of Linnæus's great works—the " Genera" and " Species Plantarum," and the " Systema Vegetabilium." The method I have hitherto pursued is this—&c. &c. . . . .

In consequence of what I had said in the Gentleman's Magazine for October last, I had a letter from Dr. Darwin, informing me that they had almost completed a Catalogue of Botanic names and terms accented, at Lichfield, which they should soon publish. Dr. Goodenough has, also, long been about a *Botanica Metrica;* which I presume will be more extensive and classical, as I understand he means to give authorities. . . . . . .

THO. MARTYN.

From Dr. Pulteney to Professor Martyn.

Blandford, Aug. 31, 1786.

Dear Sir,

I thank you for your kind remembrance of me by Mr. Hudson. . . . . He stayed with me four days, during which time I had an opportunity of carrying him with me towards Christ Church; when we spent an hour or two in *simpling* on the heath; but the season was too late, and I had only an opportunity of showing him the *Utricularia vulgaris,* [*Greater Bladder-wort,*] and a few other of the rarer plants, in the ditches near the river Avon. . . . . I most heartily

wish you health and strength to conduct to your
wishes the great work you have undertaken. I do
not see that you can improve your plan for the public
in general. . . . . No one can have felt more disgust
and aversion in the exercise of physic, as an *art* and
a *trade*, than I have done, delightful as the *study* of it
may have been; and I have a thousand times wished
for the privilege of giving myself up to Botany; *sed
stat sua cuique dies !* . . . .

RICHARD PULTENEY.

In 1786, Professor Martyn was elected a Fellow of
the Royal Society.   The next year he published,
without his name, 1. *" The Gentleman's Guide in his
Tour through Italy; with a correct Map, and directions for travelling in that Country.   London, 1787."*
12mo. pp. xlvi. and 398.—2. *" An Appendix to the
Gentleman's Tour through Italy; containing Catalogues of the Paintings, Statues, Busts, &c. London.
1787."* 12mo. pp. 158.—3. *" Sketch of a Tour through
Switzerland; with an accurate Map. London, 1787.'*
12mo. pp. iii. and 95.—4. *An Appendix to the Sketch
of a Tour through Switzerland; containing a short
account of an Expedition to the summit of Mont Blanc,
by M. de Saussure, of Geneva. London. 1788."* 12mo.
pp. 32, [numbered from 96 to 127.]—The Tour
in Italy was afterwards republished with very considerable additions, under the following title :—
*" A Tour through Italy; containing full Directions
for Travelling in that interesting Country; with ample*

*Catalogues of every thing that is curious in Architec-
ture, Painting, Sculpture, &c., some Observations on
the Natural History, and very particular Descriptions of
the four principal Cities —Rome, Florence, Naples, and
Venice, with their Environs.   With a coloured Chart:
by* Thomas Martyn, B. D., F. R. S., *Professor of
Botany in the University of Cambridge. London,* 1791."
8vo., pp. xxix. and 480.   The observations on
Natural History, even in this enlarged edition, are
very scanty; a circumstance much to be regretted.
" I wish," he writes to Dr. Pulteney, (May 9th,
1791,) " my book were more to *your* purpose and
mine; but my stay in Italy was too short, and I was
too much engaged in other pursuits to glean any
·Botanical knowledge worth communicating to the
public."—The volume is to be regarded simply as a
traveller's Guide; in which view it was considered
very useful, but it is now superseded by more modern
and complete tours.

From Professor Martyn to Dr. Pulteney.

Park Prospect, March 5, 1787.
Dear Sir,
. . . . The Book which I now send you, though
a very small one, has taken a good deal of time both
to write and print. . . . . Now that this *little* affair
is done with, I shall attend entirely to the *great* one,
in which I have made no progress this winter.   I am
pleased that our stars consent so wonderfully; and I

approve so entirely of the hints that you have given me, as to hope for the favour of more when they come in your way.   I shall *accent* the generical terms ; . . . and give the etymology of each, also, where it is known to me ; and, as you advise, dedicate a few lines to the memory of those Botanists whose names the plants bear ; but the notices must be short, otherwise the work will be too much swelled by what after all is extraneous matter.   I shall, also, gladly attend to the other hint you give ; of adding the time when, and the persons by whom, foreign curious plants have been first introduced: it will certainly be an interesting topic of information, and it is a debt of gratitude to such as have perhaps risked their health and lives to add to our knowledge, comfort, or pleasure. . . .

Mr. Curtis has started a new work, which he calls a *Botanical Magazine,* which is to exhibit the plants generally cultivated in gardens ; so that, between his *Flora* [*Londinensis*] and this, if he lives to the age of Methuselah, we shall have coloured figures of most plants commonly known.   But, whatever *he* may do I do not think that either you or I shall live to see an end of these two works. . . . .

THO. MARTYN.

About this time he contributed several notices of the natural history of Switzerland, to Mr. Coxe's Travels in that country, the first edition of which was published in 1789.   Dr. Pulteney, also, rendered

N

Mr. Coxe considerable assistance in the same department.

From Professor Martyn to Dr. Pulteney.

                                   Park Prospect, Nov. 8, 1787.
Dear Sir,

When I had the pleasure of receiving your last kind letter, of July 6th, I was on the eve of going to Cambridge; and since my return, I have been so hard at work upon the *Dictionary*, that I have much neglected my correspondents. My zeal begins now to relax again, and I feel myself more willing to attend to other matters. I indulge myself in these deviations from the main point, in order to return to it again with more vigour. My business at Cambridge was to set my Museum again in order, in the new apartment, built for me by the University, that every thing may be ready for my Lectures next spring. . . .

I had the pleasure of seeing Mr. Hudson before he left town. I sincerely hope you may send him back to us from the West in better health.

I was very glad to find that Mr. Coxe's *Faunula Helvetica* was in such good hands. He was very full of gratitude for your great kindness to him, when I last had the pleasure of seeing him. Nobody is more active than he, and I dare say, the public will yet be indebted to him, for many more useful works. . . .

                                          THO. MARTYN.

From Professor Martyn to Dr. Pulteney.

Park Prospect, Jan. 1, 1788.
Dear Sir,

Your favour of Dec. 24, contains exactly the information I wanted. The *Dictionary*, however, is in danger of becoming so big, that I must not be too diffusive in an article* which, though important, is certainly rather out of the way. . . . The labours of twenty men for twenty years, would hardly suffice to make the work I have undertaken a *complete* one ; and, therefore, how far short of perfection it must needs fall in *my* hands is obvious :—however, all candid persons, who are acquainted with the subject, will make proper allowances, and receive what is done with gratitude. . . . .

Many years ago I applied myself much to English Botanical Biography, and collected many of our antient authors, with the view of publishing a book on *the Rise and Progress of Botany in England, from the earliest time to the institution of the Royal Society.* But that, like many other projects, soon vanished in smoke, and was thought of no more. . . . .

TheO. MARTYN.

A patron, who was wholly unknown to Professor Martyn, nominated him to the Donative of Edgware,

* [The medicinal and economical uses of plants.]

N 2

in Middlesex, on the 18th of January, 1788. This
unexpected gift was bestowed by William Lee An-
thonie, Esq., at the suggestion (as he had reason to
believe) of Sir William Lee, Bart., a relative of the
patron, who was also personally unknown to him ;
but who was well acquainted with his character and
merits. On the 15th of July, in this year, he was
admitted a Fellow of the Linnæan Society, recently
instituted ; he was afterwards elected one of the Vice-
Presidents, in which office he continued as long as he
resided in London.

In the autumn of this year he issued a Prospectus
of his plan for his " *Dictionary*," and solicited the
assistance of scientific correspondents, through the
medium of the Gentleman's Magazine. (Gent. Mag.
1788, Part II., p. 867.) But this public appeal produced
him no literary aid.

About this time some communications took place
between him and the poet Cowper ; with whom,
however, he never had the privilege or opportunity of
forming a *personal* acquaintance, though invited to do
so. The short and little intercourse which occurred,
was carried on through the medium of an amiable
friend, and distant relation of the Professor,—the late
Mrs. King, wife of the Rev. John King, Rector of
Pertenhall, in Bedfordshire. This lady corresponded
freely with the endeared bard, between the years
1788—1792 ; many of his letters to her were furnished by Professor Martyn to Dr. Johnson, of Yax-
ham, for his interesting publication, containing " The

private Correspondence of Cowper with his friends."
When the poet was busily engaged with his "Homer,"
in 1789, Professor Martyn called his attention to the
valuable remarks on *translation*, by his old College
friend, Mr. Twining, in his " Aristotle."—" Being,
and having long been, so deep in the business of
translation," (writes Cowper, on the 12th of March,
1789,—) " it was natural that I should have many
thoughts on that subject. I have, accordingly, had
as many as would of themselves, perhaps, make a
volume ; and shall be glad to compare them with
those of any writer recommended by Mr. Martyn.
When you next write to that gentleman, I beg you to
present my compliments to him, with thanks for the
mention of Mr. Twining's book."—" I have not yet"
(he adds, on the 22d of April) "had time to do justice
to a writer so sensible, elegant, and entertaining,
by a complete perusal of his work ; but I have with
pleasure sought out all those passages to which
Mr. Martyn was so good as to refer me, and am delighted to observe the exact agreement in opinion,
on the subject of translation in general, and on that
of Mr. Pope's in particular, that subsists between Mr.
Twining and myself. 'Ornament for ever!' cries
Pope.—'Simplicity for ever!' cries Homer. No two
can be more opposite."—The following letter alludes
to the great work in which Professor Martyn was
now engaged ; and in which the poet could feel an
interest, though in a department so different from his

own;—it is not, however, surprising that one who was himself so passionately fond of a "*Garden*," should duly appreciate the scientific pursuits of an eminent *Botanist!*

From William Cowper, Esq., to Mrs. King.

January 14, 1790.

My Dear Madam,

     . . . . I felt a true concern for what you told me in your last, respecting the ill state of health of your much valued friend, Mr. Martyn. You say, if I knew half his worth, I should, with you, wish his longer continuance below. Now you must understand that, ignorant as I am of Mr. Martyn, except by your report of him, I do nevertheless sincerely wish it—and that, both for your sake and my own; nor less for the sake of the public. For your sake, because you love and esteem him highly; for the sake of the public, because, should it please God to take him before he has completed *his great Botanical Work*, I suppose no other person will be able to finish it so well; and for my own sake, because I know he has a kind and favourable opinion before hand of my translation [of Homer,] and consequently, should it justify his prejudice when it appears, he will stand my friend against an army of Cambridge critics.—It would have been strange indeed if *self* had not peeped out on this subject. I beg you will present my best respects to him, and assure him, that, were it possible he could visit Weston, I should be most happy to receive him. . .

WILLIAM COWPER.

Mr. Martyn having expressed to Mrs. King his admiration of the poet's writings,—his praise called forth the subjoined remarks :—

From William Cowper, Esq., to Mrs. King.

Nov. 29, 1790.

My Dear Madam,

I value highly, as I ought, and hope that I always shall, the favourable opinion of such men as Mr. Martyn ; though, to say the truth, their commendations, instead of making me proud, have rather a tendency to humble me, conscious as I am that I am over-rated. There is an old piece of advice, given by an antient poet and satyrist, which it behoves every man, who stands well in the opinion of others, to lay up in his bosom : *Take care to be what you are reported to be.* By due attention to this wise counsel, it is possible to turn the praises of our friends to good account, and to convert that which might prove an incentive to vanity into a lesson of wisdom. I will keep your good and respectable friend's letters very safely, and restore it to you the first opportunity. I beg, my dear madam, that you

will present my best compliments to Mr. Martyn,
when you shall either see him next, or write to him.

To that gentleman's enquiries I am, doubtless, obliged for the recovery of no small proportion of my
subscription-list; for, in consequence of his application to Johnson, and very soon after it, I received
from him no fewer than 45 names, that had been
omitted in the last he sent me, and that would probably never have been thought of more. . . .

WILLIAM COWPER.

His friend, Dr. Pulteney, having transmitted to him
a work he had just published,—" Historical and Biographical Sketches of the progress of Botany in
England, from its origin to the introduction of the
Linnæan System,"—the gift was acknowledged in the
following letter :—

From Professor Martyn to Dr. Pulteney.

Park Prospect, June 14, 1790.

Dear Sir,

 . . . I have run over your two volumes with
great avidity; and poured some tears of pleasure over
your account of the excellent Ray, who might claim
a commemoration from English Naturalists, with at
least as much justice as Handel* from the Musicians;

[* Alluding to the grand Commemoration of Handel, in Westminster Abbey, in 1784, at which Professor Martyn was present.]

when to his various knowledge we add his many valuable qualities, and sincere piety. I have to thank you, particularly, for the justice and honour you have done my revered father; as well as for the manner in which you have occasionally mentioned me—particularly for speaking of that correspondence with which you have honoured me during so many years, and from which *I* have reaped so much improvement. Had I known of your design, I should probably have sent you my chaotic collections[a] on the same subject. You have, perhaps, had a fortunate escape; as the gleanings would scarcely have paid you for the trouble of looking them over. As it is, perhaps, when I have leisure to give your Sketches a more orderly reading, I may send you some observations.[b]

I have been much interrupted, during the winter, in the prosecution of the *Dictionary*, by a long illness. . . . . .

Tho. Martyn

His old friend, the Rev. G. Ashby, furnished him with several useful hints, and some little assistance in

[a] [See above, p. 179.]

[b] [Some criticisms were transmitted in the autumn of 1790; of which Dr. Pulteney intended to avail himself, had his work proceeded to a second edition, (Dr. Pulteney's Letter to Professor Martyn, 24th May, 1791.)]

the compilation of the great work alluded to at the close of the preceding letter; particularly by collecting the vulgar and trivial names of English plants, from the oldest writers. The two following letters allude to this subject.

From Professor Martyn to the Rev. G. Ashby.

Cambridge, Sidney Lodge, Sept. 1, 1790.

Dear Sir,

. . . . I thank you much for your hint about Lyte, whom I have looked over, and find fertile in English names; I wish he and others, instead of coining new names, had collected the few that were actually in use in different counties. For want of knowing these, we cannot yet talk intelligibly to peasants, or even to farmers. For instance, the *Corn Poppy** is called the *Canker* in Suffolk. I do not find this name in any of the books; whether it be in your list I know not, as I have it not with me here. If the vulgar names of weeds and plants, in common cultivation, were well collected, it would much facilitate agriculture. But here, as in almost every thing else, there has been so little connection between theoretic and practical men, that we hardly understand one another.

For book names we have a connected chain, from the *Grete Herball* by *Treveris,* in 1516, to our time;

* [*Papaver Rhœas.*]

through *Turner*, in 1551 and 1568 ; *Lobel*, in 1570 ;
Lyte, in 1578 ; *Gerard*, in 1597 ; *Lobel*, in 1605 ;
Parkinson's Paradisus, 1629 ; *Johnson's Gerard*, 1636 ;
Parkinson's Theatrum, 1640 ; *Ray*, from 1660 to 1696 ;
the third edition of his *Synopsis*, [in 1724 ;] Hudson,
[in 1762 ;] Withering, [in 1786 ;] &c. &c.   Have you
looked over all these ? If any besides Lyte are
omitted in your list, be so good as to inform me ; as
also if you recollect any others.   Probably, if one had
time, some names might be picked out of our 𝖆𝖚𝖓𝖈𝖎𝖊𝖓𝖙
𝕰𝖓𝖌𝖑𝖕𝖘𝖍𝖊, *not* Botanists ;—as from Chaucer[a], 𝕮𝖑𝖔𝖜𝖊,
𝕲𝖞𝖑𝖔𝖋𝖗𝖊, 𝖆𝖓𝖉 𝕷𝖞𝖈𝖔𝖗𝖎𝖈𝖊, &c.

But I have already worked hard on the Dictionary
five years, and have scarcely finished one third of it.
I have already undertaken too much, and if I undertake more, the work will never come before the
public. . . . .

THO. MARTYN.

From the Rev. G. Ashby to Professor Martyn.

Barrow, Sept. 4, 1790.

Dear Sir,

. . . . I am sensible what a mass of matter you
have to move ; I suppose the Booksellers may have
insisted on Miller's name being preserved ; else, I think

---

[a] [The reference is to Chaucer's *Romaunt of the Rose*. (Edit. 1542.)
   " 𝕮𝖍𝖊𝖗𝖊 𝖜𝖆𝖘 𝖙𝖍𝖊 𝖎𝖓𝖊.𝖙𝖞𝖓𝖌𝖊 𝖒𝖆𝖓𝖞 𝖆 𝖘𝖕𝖎𝖈𝖊,
   𝕬𝖘 𝕮𝖑𝖔𝖜𝖊, 𝕲𝖞𝖑𝖔𝖋𝖗𝖊, 𝖆𝖓𝖉 𝕷𝖞𝖈𝖔𝖗𝖎𝖈𝖊,"]

you could have written it all new with more ease and
credit. Miller, being designed for working gardeners,
who do not understand *feuilleting* a book, is justified
in giving the same minute directions for culture to
two articles that stand next to one another. Another
excuse for him is, that his book at first was hardly one
sixth so big as it since grew to. I have sometimes
thought that a vast reduction might be accomplished
by adding one more article—CULTURE. Specifying as
many sorts as you can, call them 1, 2, 3, 4, 5, 6, &c.;
and, as you have occasion, say, Give Culture No. 1,
or No. 6, as proper. After all that can be done, the
vast size of the volume seems to make it unfit for
actual use; it cannot be carried into a garden, and is
too costly to be thrown about. . . . .

. . . . You are quite right as to English names. I
have read all old *Bullein's** works, and have been
astonished at his names, which no books will help you
to understand; and those so misprinted, where health
and life are concerned, that one can wonder at nothing
of this kind in W. Shakspeare. (By the bye, is it not
wonderful that neither the 1st nor 2d editions of the
god of our idolatry is in any of the Libraries of Cam-

* [The curious works of William Bullein, are as follows:—1. A
newe Booke, entituled The Gouernemente of Healthe. 12mo.,
1558.—2. A comfortable regimen against the Pleurisie. 12mo.,
1562.—3. A Dialogue, both pleasaunt and pietifull against the
Feuer-Pestilence. 8vo., 1573.—4. Bulwarke of Defence, against
all sicknesse, soarnesse, and wonndes, that doe daily assaulte
Mankinde. Fol. 1579.]

bridge, that I can get at.)—For *weeds*, you will find good help in the provincial Glossaries to Mr. Marshal's Agriculture. He wants little of being excellent; perhaps mostly in method :—how troublesome is it to make out the history of the Turnip-fly, or Cheese-making, in his diurnal relation. Was there ever such a lucrative project—(may I call it *hum ?*)—as, for saving Turnips from the fly by night-rolling! He proposed receiving 2000 guineas for the discovery, and had the modesty to publish it when he got only 1500. Commend me, also, to the man who has plagued Parliament for a reward for an insect-killing powder! I believe he expects £5000 or £6000, on the usual pretence of having made the discovery after great loss of time and money. As to the efficacy of his secret, take the following, which you may depend upon. He applied to my neighbour, H. Mure, Esq., (from whom I had it immediately,) for his signature to his petition. He said,—" Let me first see what it will do: there is a nest of ants, that have made a road over the gravel walk, strew your powder there." He did so, as plentifully as he pleased; and the consequence was, that the ants deserted *that* track, and took *another* a very few inches to the right or left.— *Cankerweed*, in Marshal, is *Senecio Jacobœa*, or Common Ragwort.—Ray I have deflowered, I believe accurately. Hudson has very few, if any, exclusive of his own make. . . . .

G. Ashby.

From Professor Martyn to Dr. Pulteney.

Park Prospect, May 9, 1791.

Dear Sir,

I know not whether you see the Reviews, or think it worth while to attend to what they say. I have just slightly looked over what the Monthly Review says of your work. The general account is handsome; as who, indeed, can treat it any otherwise: but some fault is found with it, in my opinion not with any great judgment. Among other things, I observe that the Reviewer will not allow your Epocha of the establishment of Botany in England; but would have it date from the reception of the Linnæan names by the College of Physicians[*]! Though it may be allowed

---

[*] [The strange passage to which Professor Martyn alludes, is in the Monthly Review for 1791, Vol. IV. pp. 367—369. " Botany speaks principally to the salutary purposes of the *Art of Healing;* of course the concurrence of those who *preside* over the Art of Healing, gives the most authentic countenance to the *Medicinal modes* which are most deserving of attention. How divided would the Botanical Empire have still remained, if the London College had still upheld the nomenclature of the ancient Authors !"—May we not rather say, How slow would be the progress of human knowledge, if Science were thus to await the *Cum Privilegio* of the Royal College of Physicians, or of any other learned body which might vainly arrogate to itself the power of giving its authoritative stamp to philosophical systems !]

that the *Lectures* in two of our Universities can hardly
form an Epocha; yet, certainly, they were the first
public notices of the Linnæan system, and have contributed much to spread it.   At my first Course, in
1762, [1763 ?,] about a hundred pupils attended; and
among them many who were persons of the first
figure for literature, and others who have since been
not inferior to them.   In my different Courses, I suppose upward of 500 persons have been instructed;
and they have been dispersed all over the kingdom:
they must, therefore, in some degree have disseminated
the knowledge of Botany, and of the Linnæan System
among us.   Dr. Hope having a larger class, and having taken more pains on the subject, must have done
much more.   Other causes, probably, co-operated,
particularly in the Metropolis: but I well remember
that Botany was then at a low ebb; and even ridiculed
abundantly, along with the study of Grubs and Cockle-
shells!   Such Botanists as yourself, and Sir William
Watson, were not very numerous.

THO. MARTYN.

From Dr. Pulteney to Professor Martyn.

Blandford, May 30, 1791.

Dear Sir,

. . . . The writer in the Monthly Review,
seems not sufficiently aware, that I meant studiously

to avoid coming down, in *detail*, among living authors. In fact, I came lower than, perhaps, I ought; as from the title I was not obliged to come so low as to the *establishment* of Linnæan Botany in England, but *only* to its '*introduction*,' which was twenty years or more, before I fix my æra of *establishment*. I had the *Genera Plantarum*, as early as 1749, in my possession; and I am by no means convinced, that I fixed upon a wrong æra, for the *establishment* of it. Your last letter agreeably confirms this opinion. I guess the writer of the Article in the Monthly Review not to be a real Botanist. . . .

R. PULTENEY.

From Professor Martyn to Dr. Pulteney.

Park Prospect, June 7, 1791.

Dear Sir,

. . . . I am glad we agree so well on the establishment of Botany. I possess the first edition of the " *Genera Plantarum*," and also the " *Critica Botanica*," [published in 1737,] presentation books to my father, with " *Clarissimo Prof. Martyno*," in Linnæus's own hand.

I am getting on with the *Dictionary*, and have completed letter F; but, as I mean it to be " a digest of the Botany of the age," (—to use one of your own expressions—) it is a most laborious business; and such works are daily coming out that I

am obliged to go over the ground again and againe
and, after all, it must remain very imperfect.   I hav;
solicited assistance, but almost in vain!

THO. MARTYN.

Oppressed as he was with the burden of this great
work, it is almost beyond belief that this indefatigable Botanist should have had the courage to allow
his attention to be divided between his principal
object, and any collateral occupations; nevertheless,
at this very time, his mind, and his pen were engaged on *other* scientific subjects, which might well
have engrossed *all* the leisure of a writer of ordinary
perseverance.   Some of these come naturally under
our notice in this place.

A most important subject, connected with the modern improvements in the science of Botany, presented itself to his view, almost as soon as he had
fairly set to work upon his *Dictionary*.   This was, to
give to Botanical *language* the same accuracy and
precision in the *English* tongue, which it had received
in the *Latin* from the philosophical genius of Linnæus.   The *Philosophia Botanica* of Linnæus, in 1751,
laid the foundation of a definite and concise language
in this science.   But the first complete list of Botanical terms, accompanied with explanations, and
detached from all other matter, appeared in the sixth
volume of his *Amœnitates Academicæ*, in 1764; it is
entitled *Termini Botanici*, and was a Thesis, read by
J. Elmgren, in 1762. — Professor Martyn, having

o

determined to desert the learned language, in which
the technical descriptions of plants had hitherto (with
few exceptions) been conveyed, and having resolved
to communicate the treasures of Botanical knowledge
to his countrymen through the medium of their
native tongue; it was manifestly a matter of the first
importance, to fix, with judgment, the *English termi-
nology* of the science. His first thoughts on this
subject were communicated to the Linnæan Society,
in the following paper,—" *Observations on the Language of Botany; by the* Rev. THOMAS MARTYN, B.D.;
F. R.S., F.L.S., *Professor of Botany in the University
of Cambridge. Read,* Oct. 6, 1789." (Trans. of Linnæan Soc. for 1791, Vol. I. pp. 147—154.) In this
paper he observes,—that " the excellent language
which Linnæus invented will continue to be in use,
even though his system should, in after ages, be neglected, and will be received into every country where
the science of Botany is studied, with certain modifi-
cations, adapting it respectively to each vernacular
tongue. So long as Botany was confined to the
learned few, there was no difficulty in using the terms
of the Linnæan language, exactly as the author had
delivered it; but now that it has become a general
pursuit, not only of the scholar, but of such as have
not had what is called a learned education; and since
the fair sex have adopted it as a favourite amusement; it is become necessary to have a language
that shall be suitable to every rank and condition;
a language that may be incorporated into the general

fund, and carry with it the proper marks of the
mother tongue, into which it is to be received."—He
objects, however, to abandoning the old scientific
terms, for those which, though purely English, need
quite as much explanation to the unlearned; thus,
he prefers the use of " *calyx*" rather than " *empale-
ment.*" On the other hand, he protests against the
adoption of a learned or pedantic term, when a long
established or significant English word exists; thus,
he would write " *seed-vessel,*" rather than " *pericarp;*"
and " *bell-shaped,*" instead of " *campaniform.*" On
this last sensible observation, he enlarges, as follows:—
" When we admit terms of art, to participate in the
rights of citizens, they should put on *our* garb, and
adopt *our* manners. If this rule had always been
observed, our language would not have been de-
formed with innumerable barbarisms, which learned
and unlearned ignorance have joined to introduce
among us; and which nothing but the constant habit
of hearing or speaking them, could ever reconcile to
our ears. . . . It should seem, that the mercantile
world, the learned world, and the fashionable world,
had formed a conspiracy to debase our sterling English,
by ill-made terms, affectedly introduced without the
least necessity."

This paper laid the foundation of a regular treatise
on the subject, which has been highly approved by
the scientific world, and which has afforded useful
materials to later writers,—indeed, it is by no means,
itself, an obsolete work.—" *The Language of Botany,*

*being a Dictionary of the terms made use of in that science,
principally by Linnæus : with familiar explanations, and
an attempt to establish significant English terms. The
whole interspersed with critical remarks. By* THOMAS
MARTYN, B. D., F. R. S., *Professor of Botany in the
University of Cambridge. London,* 1793." 12mo. A
second edition was published in 1796, and a third,
corrected and enlarged, in 8vo., in 1807. An extract*
from the preface, will explain the principles of this
excellent work, in the author's own interesting manner.—He begins by stating the progress of his own
experience, with regard to the value of precision in
Botanical language :—

" My attention was first called to consider the language of Botany, very soon after Linnæus had published his fundamental treatise, (in the year 1751).
At that time I was a pupil in the school of our great
countryman, Ray. But the rich vein of knowledge,
the profoundness and precision which I remarked
every where in the *Philosophia Botanica,* withdrew me
from my first master ; and I became a decided convert to that system of Botany, which has been since
generally received. . . .

" Having been appointed by the unanimous voice
of the University of Cambridge to the Professorship of
Botany . . . I had the felicity in taking the lead in
introducing the Linnæan system and language to

* [A small portion of this extract has already been quoted
above, p. 101 ; but it was necessary to repeat it here, in order
to connect the passages which follow.]

my countrymen, by a course of public lectures (in 1762*). . . . .

" The institution of the Linnæan Society; the avidity with which the study of Botany has been lately pursued by many in every rank and description of persons; the necessity I was under to find terms by which to express myself in my *Letters on Botany,* and especially in the great work which I am now finishing [*Miller's Gardeners' and Botanists' Dictionary, &c. &c.*]; have all conspired to excite my attention to Botanical language, and particularly to the mode which seems best for us to adopt when we write or speak of the science in our native tongue.

" So long as Botany continued to be studied only among those who had received a learned education, the original terms of Linnæus, derived from the Greek or Latin, served all the purposes of general intercourse. But when it became universally adopted, a vernacular language would of course be gradually formed; and, if it were to be left to chance, or the choice of the ignorant, many absurdities and barbarisms would be introduced, debasing our sterling English. This it has been my wish to avoid; and I now renew the attempt which I made some time since (*Dissertation,* printed in Vol. I. 147-154, of the Transact. of the Linnæan Soc.) to fix our native Botanical language on certain and reasonable principles, conformable to general analogy. Had not this been my particular view, and had I been

* Query, 1763 ? See above p. 116.

satisfied with what has been already done by several
learned and ingenious writers, I should certainly not
have obtruded my ideas upon the public, after such a
multitude of elementary books had been printed;
and even now, the errors, omissions, and defects of
various kinds, which those who are skilled in philolo-
gical Botany will easily detect in this little volume,
require an apology. I must request the public, therefore, to consider it as a mere attempt, that may hereafter be improved into something more worthy of
their regard, if learned Botanists and Philologists will
condescend to consider the subject more deeply.

" I am aware that many will say, You give too
much importance to these laborious trifles. But, if
they be such, they lead not to any serious mischief;
and so long as the weightier matters of science are
not neglected, there can be no harm in working up
and polishing the minuter parts, so that the ornaments
may not disgrace the edifice.

" The indolent, I am sensible, will shrink from this
odious assemblage of terms : but the indolent must
be contented to lie under the disgrace of ignorance, or
at most to skim very lightly the surface of knowledge.

" Many terms are indispensably necessary in the
science of Nature, where the objects that present
themselves to our consideration are so numerous.
The question, therefore, is not, whether we shall have
terms or no ; but, in what manner they should be con-
structed, so as to answer the great purpose of receiv-

ing and communicating knowledge most effectually?
Now we have been long in possession of a precise and
significant language, invented by Linnæus, generally
adopted by the learned of every country in Europe,
and received in great part into the vernacular tongues
of several.   Can we do better, therefore, than to keep
as close as possible to this, and to adopt the Linnæan
terms themselves, so far as the nature and structure of
the English language will permit, and whenever we
can do it without violating the laws of grammar or
common sense?   We shall thus have all the advantage which is derived from speaking and writing one
universal language; whereas, if we set about finding
equivalent terms in English, these will require as much
explanation as the others, and will be equally difficult
to the student, without having possession or prescription to plead.   Thus shall we become unintelligible
to every other nation, without being more intelligible
among ourselves.

" Laying it down, therefore, as a first principle,
that we ought to adhere as closely as possible to the
Linnæan language, it will be found that the number
of terms purely English, occurring in the Botanical
Glossary which is now offered to the public, is comparatively small. . . . .

" That we must depart sometimes from the Linnæan language, I readily allow; but the cases are
rare, and the instances under each case are not many.
. . . . These exceptions being admitted, I hope to be
excused for repeating my opinion—that the advantage

of Botany will most effectually be consulted by retain-
ing the Linnæan terms, whenever there is no cogent
reason to the contrary."—

· Another work, which much occupied his attention
about this period, was—" *Flora Rustica: exhibiting
accurate figures of such plants as are either useful or
injurious in Husbandry.* Drawn and engraved by
FREDERICK P. NODDER, *Botanic Painter to her Majesty,
and coloured under his inspection.* With scientific
characters, popular descriptions, and useful observations:
by THOMAS MARTYN, B. D., F. R. S., F. L. S., *Professor of Botany in the University of Cambridge.* London, 1792—1794." 4 vols., 8vo. This work was
begun in the autumn of 1791, and published in numbers. It is dedicated to the King. It was Professor
Martyn's design to " present the public with figures
and descriptions of those plants with which the
husbandman is principally concerned;" especially *the
Grasses,* the whole of which he intended to have included, had the work met with sufficient support.
Only 144 plants, however, were given ; the pecuniary
encouragement not being such as to warrant the Authors in carrying on their plan further. The work, as
it is, exhibits about thirty plants useful in culture or
the arts (exclusive of the Grasses, Trefoils, and
other Legumes) ;—forty species of Grasses ;—fourteen
of the Trefoils ;—about twenty-seven of the weeds
infesting pastures ;—and upwards of twenty of those
which are hurtful to arable lands. This interesting
and valuable work, never having been reprinted, is

very scarce, and fetches a high price. It is not un-
common, however, to meet with cheap copies which
profess to be perfect :—these spurious copies bear the
name of a bookseller of the name of Harding, of St.
James's Place, at the foot of the false title-page ; and
have been formed from the waste-sheets of the original
work, completed (after a manner) by the insertion of
many leaves miserably re-printed on bad paper ; while
the plates are worn-out impressions, and *not* coloured !

Besides these laborious undertakings, immediately
connected with the Professor's Botanical pursuits, another engagement, in a very different department of science, devolved upon him about this period; this was the
office of Secretary to the Society for the improvement
of Naval Architecture, established in 1791.   This Society owed its origin to Mr. Sewell, the editor of the European Magazine ; who, having been induced to make
frequent visits to the coast for his health, was much
struck with observing that Naval Architecture was
almost neglected as a science, and that almost all
the improvements of ship-building were derived from
the French.   Mr. Sewell called the attention of his
countrymen to this subject in 1791, in the periodical
work of which he was the Editor; and soon after published a volume entitled, " A Collection of Papers on
Naval Architecture, originally communicated through
the channel of the European Magazine ; Part I., London, 1791."   8vo.   Professor Martyn's attention was
called to this subject about the same time, by his
friend, Sir J. B. Warren ; and through the exertions of

these gentlemen, combined with those of Mr. Sewell,
the above-mentioned Society was formed in the spring
of 1791.   The following letter will further explain
the circumstances of its establishment.

From Professor Martyn to the Rev. G. Ashby.

Park Prospect, Dec. 12, 1791.

My dear Sir,

As the Society for the improvement of Naval
Architecture may hereafter be of consequence to this
Country, you will not be displeased to know how it
originated.

My friend, Sir John Borlase Warren, who is a
Captain in the Navy, desired me to go with him to
Sewell, the bookseller in Cornhill, to settle matters
for printing a book which he had composed, and has
since published, under the title of—" *A View of the
Marine Force of Great Britain.*" Our conversation
turned upon naval affairs, and the great ignorance
that prevailed in the principles of ship-building.   Mr.
Sewell said that he had been thinking of the subject
for thirty-five years; upon which my friend observed,
that he had *thought* upon the subject long enough, and
that it was now time to *do* something.   We agreed
that so considerable a national object would probably
be taken up with ardour, if it was once laid before the
public.   I promised to draw up a proposal, which I
did immediately; and, having showed it to my friend
and Sewell, together with a Mr. Rogers, a watch-

maker in the city, an intimate friend of Mr. Sewell's,
we printed and dispersed it directly.   This produced
a meeting; another soon after on the 20th of April;
a numerous list of subscribers; and respectable patronage. Thus has a great Society for naval improvements been formed by a bookseller, a watch-maker,
and a parson, with the assistance of one who is rather
more a fashionable man than a naval officer. After
a few meetings I drew up the address, and the rules
and orders for our government.   I have continued
ever since to conduct the whole business of the
Society as their Secretary. We have offered some
handsome premiums; and we are soon to try some
experiments on a very large scale, at the expense
of the Society.

On the first of November last, I launched a very
small vessel, under the name of *Flora Rustica* ; which
perhaps may not have reached your ears. To drop
the metaphor,—it is a periodical publication, to contain
three coloured plates in each number, of grasses and
other plants useful or destructive in husbandry.

The great work goes on, and I have promised to
encounter the printers in the spring. . . . .

I observe that the trees grow old very fast in St.
James's Park, and I fancy the case is the same with
the human species in this region of fog and smoke.
There are scarcely twenty of the original Elms left,
and *they* look most deplorably. . . . .

THO. MARTYN.

He continued to conduct the correspondence of
the Society for Naval Architecture, during three
years ; but was ultimately obliged to resign an office,
the duties of which he found incompatible with his
other pursuits.   On the 8th of April, 1794, he was
presented with the Society's first gold medal, as an
acknowledgment of his zeal in its original institution,
and of his valuable services in the conduct of its
affairs.   This Society published some valuable papers,
particularly those on the resistance of fluids.   It was
dissolved about 1796, in consequence of the want of
funds ; indeed, it contracted a debt which was paid
chiefly by Earl Stanhope, Admiral Hamilton, and
Colonel Beaufoy.   Its objects are now fully answered
by the Naval College at Portsmouth, (which embraces
a School of Naval Architecture,) and by the spirited
investigations of individual experimentalists.   The
following extract from a letter, addressed to him about
this time by his ingenious friend, the Rev. George
Ashby*, of Barrow, (while playfully alluding to Professor Martyn's recent honours in having obtained
the gold medal of the Society), contains also some
original and not altogether unjust remarks on the
wide difference between ancient and modern medals,
as commemorative of merit.—" Large medals are

---

* A curious and intelligent letter, written to Professor Martyn by
Mr. Ashby, Sept. 3, 1793, exists among the Banksian MSS. in
the British Museum. The subject is, the formation of harbours,
and the stability of vessels.

not the thing : they are expensive in the die, and *that* is liable to break. And, indeed, all medals, on the *modern* scheme, are but simple things. The gold, as to value, is not better than copper, for you have no thoughts of melting it : this, indeed, might be remedied, by giving it in copper along with specie ad valorem ! Ancient Medals,—from the quantity of the same in *circulation*,—served formerly, and continue to this day, to *notify* an action or person ; but how doth this *solitary* one, which you will lock up in your bureau as soon as you have got it ? As some help, let me advise you to recommend to the Society to have a good engraving of it, and print it in *every* title page of *every* even the smallest tract you publish. When Dr. Powell first instituted examinations at St. John's, he commissioned me to inquire about a die for a medal. I did so, but told him, as above, that medals had much changed their nature ; for who then knew that Craven* had got two or three ? I therefore recommended it to him to give a silver standish ; which he readily adopted. This, you will observe, though not quite the same as an *ancient coin,* has several advantages over a *modern medal :* it is not, indeed, *in circulation ;* but it may always stand *on the table ;* and the poorer the accompanying furniture is, the more it attracts the visitor's notice,—who naturally begins to examine, and then the inscription tells the whole

*[Dr. Craven succeeded Dr. Powel as Master of St. John's College. Mr. Ashby had been *President* of St. John's, before he was presented to the living of Barrow, in Suffolk.]

story, which locked up medals never can do, *avec bienseance.* However, I am glad you are to have one."—

Having given this general account of the various literary and scientific pursuits, which engaged Professor Martyn's attention, at the same period in which he was busily preparing for the publication of the first part of his great work,—"The Gardener's and Botanist's Dictionary,"—we shall now return to the year 1792, and present the reader with some further extracts from his correspondence.

From Professor Martyn to Miss Hawkins.

Park Prospect. *Not dated.* [August ? 1792.]

My Dear Madam,

I cannot doubt but that you will like the neighbourhood of Croydon. Perhaps I think so, because I have liked it myself. The soil and air are certainly good, and the water cannot fail of being so in chalk. I should not suspect the air of being keen; but I inhaled it at a time of life when my constitution was not so tender as it is now.

My *Dictionary* is advanced as far as HEM; and this puts me in mind of our friend Miss Welch, who informed me so kindly on the subject of HEMERO-CALLIS, which I have just been writing about.

Professor Bradley [*Philosophical account of the works of nature,* p. 79,] says, that ' common plants often have the colour of their leaves and flowers changed,

as the nature of the soil directs.'—This we know; but
he gives a remarkable instance, which one should
like to verify.—'An instance of this was in some
roots of the double blue *Hepatica*, [*Anemone Hepatica*,] that were sent to Mr. Harrison, of Henley-
upon-Thames, from Mr. Keys's garden, in Tuttle-
fields, whose soil was so different from the ground
they were planted in at Henley, that, when they came
to blossom there, they produced *white* flowers, and
were therefore returned back to their first station
where they retook the *blue* colour they had at first.'
—The soil about Henley is all chalk. When you go to
Croydon you may set up a manufacture of *whiting*
all *blue* flowers. Now, if we could bring back Miss
Welch's *Hemerocallis*, or Day-Lilly, to its original colour, by a re-removal, it would be curious! Mention
the scheme to that good lady when you write. . .

THO. MARTYN.

The experiment to which the preceding letter alludes, was as follows. In the year 1788, Miss Welch[*]
laid out a new garden at her residence at Ardenham
Hill, near Aylesbury; into which she removed several
plants from Hampstead. Dividing a root of *Hemerocallis fulva*, the Copper-coloured Day-Lilly, she put
one part into an argillaceous soil, and the other into a

[*] "Miss Welch was a woman of distinguished mental ability;
and in great intimacy with Dr. Johnson, as her father (a well
known magistrate) had been. Her sister married Nollekins, the
sculptor."—*Note by Miss Hawkins, to the above letter.*

soil composed of rubbish from old ruins &c. The former remained unchanged; but the latter blossomed with the tint and appearance of *Hemerocallis flava,* the Yellow Day-Lilly. She communicated this curious fact to Professor Martyn; by whom it was transmitted to the Linnæan Society. (See Trans. Linn. Soc. for 1790. Vol. II., p. 353.) It is the more remarkable, because Linnæus, in his *Systema Vegetabilium,* has described these two species by their *colour,* a distinction to which he very rarely has recourse; but, in his *Species Plantarum,* he had originally considered them as merely varieties, which he named *Hemerocallis Lilio-Asphodelus* a, et β. Professor Wildenow, in his *Species Plantarum,* considers them as distinct; and yet admits that he has not discovered the permanent difference; he refers, moreover, to Professor Martyn's communication of the above experiment in the Linnæan Transactions, in proof that the one passes into the other species.

The following letter contains an inquiry, with reference to the article (in the Gardener's Dictionary,) *Ficus Carica,* which he was then preparing :—

From Professor Martyn to Miss Hawkins.

Park Prospect, Oct. 8, 1792.

My Dear Madam,

It is said, (in Grose's Antiquities,) that there is a Fig, of the white sort, confidently asserted to have been planted by Archbishop Cranmer, at Mitcham; in the garden of the Manor House, formerly the private estate of the Archbishop, and now belonging to one of his descendants. Its branches are very low; but its stem, which measures 30 inches in girth, has every mark of great age.—Will you be so obliging as to enquire, whether this venerable archiepiscopal Fig-tree yet subsists? And, if so,—whether the above account of Grose's be exact. Your friend, the Vicar, will assist you in this enquiry; but I must beg of you to see the tree with your own eyes.

THO. MARTYN.

The result of this inquiry was,—that this venerable tree had been destroyed several years before the date of the above letter.

For nearly half of this year his health was so bad, that he was incapable of much literary labour. In the autumn he was compelled, however unfit for the burden, again to apply himself closely to his task; the calls for the Dictionary were such, that the booksellers urged him to prepare for the press in the following spring:—"It must come forth," he writes (Oct. 29, 1792) to Dr. Pulteney, "with its manifold imperfections on its head; but the candid will pardon them, in a work which those who best understand the science know not to be capable of perfection." The first number was not *published*, however, till two years and a half later than this period.

From Professor Martyn to Dr. Pulteney.

Park Prospect, March 12, 1793.

Dear Sir,

You give me great comfort, under the oppression of my laborious work, by your very kind, but, I fear, too partial approbation of its plan and arrangement. The execution will call for all your candour and indulgence; but I have no fear of *such* critics, who know too well the difficulty and labour of large publications not to make considerable allowances. I much approve your hint of distinguishing the *merit* of the figures; but, alas! this, among many other useful points of instruction, must be omitted for want of time. I wish you had suggested *more* hints, and had *blamed* rather than *praised*. I should be ashamed of myself if I were insensible to the *praise* of such men as Dr. Pulteney; but I can forego the pleasure of this, for the advantage to be derived from candid *criticism*. . . . I do assure you that I have so accustomed myself to the lash, that I can bear it with great complacency, except it be laid on with ill manners or ill humour,—which I am certain cannot be done by *you*.

I think so entirely with you on the subject of *arrangement*, that, if I were now to begin the work, it should certainly appear in the order of Linnæus's system, with an alphabetical index, English and Latin : but I was to build on Miller's foundation. I almost hope that somebody will knock Miller and me on the head, hereafter, by such a publication. . . .

                                   THO. MARTYN.

The plan adverted to in the last passage is now [1829] about to be realised, in some degree, by a republication of this useful work, in a *systematical* form, following, however, the *natural* orders, with a Linnæan and alphabetical Index.   It is to be greatly condensed, to bring it into the compass of four 4to. volumes ; but it will embrace the vast additions made to Botany since Professor Martyn's edition appeared, (1807.)

Under the year 1771 (see p. 140) was mentioned an application to government, for an endowment for the Botanical Professorship ; which was liberally promoted by the Duke of Grafton, but failed, through that nobleman's retirement from the ministry.— After a delay of twenty-two years, this long protracted act of the royal bounty was exercised, in favour of Professor Martyn individually, and not in the way of a perpetual foundation.   In Lord North's administration, through the kind influence of the Earl of Carendon, a patent was obtained, on the 2d of August, 1793, by which Mr. Martyn was appointed (after 31 years' service without salary) " The King's Professor or Reader in Botany at Cambridge," with a pension of 100*l.* per annum.   A little before

Christmas, in that year, he removed from Westminster, to Frith-street, Soho.

Mr. Nichols being at this time engaged in bringing
out his magnificent work, " The History of Leicestershire," Professor Martyn furnished him (in 1793)
with " *An Account of the Natural History of Little
Dalby;*" a parish with which he became intimately
acquainted, by repeated visits, and a long residence
at Dalby-Hall, the seat of his pupil, Mr. Hartopp.
This paper is printed in Nichols's Leicestershire, Vol.
II., Part I., pp, 160—162. As it contains the only
authentic account of the origin of the celebrated
Stilton cheese, the following extract may not be un-
acceptable to the curious reader.—" Little Dalby,"
observes Professor Martyn, " is remarkable for having
first made the best cheese, perhaps, in the world ;
commonly known by the name of ' *Stilton cheese,*'
from its having been originally bought up, and made
known, by Cowper Thornhill, the landlord of the
Bell Inn at Stilton. It began to be made here by
Mrs. Orton, about the year 1730, in small quanti-
ties. . . . In 1756, it was made only by three persons.
It is extremely rich, because they mix among the new
milk as much cream as it will bear."—From a note to
this passage by the Rev. George Ashby, it appears
that Lady Beaumont was the real discoverer of this
luxury; and that Mrs. Orton, who was a servant to
an ancestor of Mr. Ashby's at Quendon, was accustomed to see it made there for family use, from a receipt furnished by her ladyship;—and afterwards

marrying, and settling at Dalby, she began to make *'Lady Beaumont's cheese'* for sale.   In an unpublished letter to Professor Martyn, Mr. Ashby adds,— "I have the receipt as practised at Quenby, when there was no such thing as a public sale; it was my grandmother's maid who married away and first sold; if I am rightly informed, her maiden name was Scarborough.   I have no doubt of the receipt coming from Lady Beaumont."

The following letter, from the same ingenious correspondent, contains a curious anecdote on the poisonous qualities of the Potato-Apple :—

From the Rev. G. Ashby to Professor Martyn.

Bury St. Edmunds, May 10, 1795.

Dear Sir,

.... "A noxious quality seems to lurk in Potatos, and connects them with their congeners. Three or four years ago I was told, that the Apples of Potatos made good tarts; and so my sisters and I found them.   But one morning a neighbouring clergyman calling on me, I told him that the oven was drawn, &c., &c., and for the novelty he was willing to taste them.   Scarcely had he put the first bit to his mouth, but he clapped his hand to his throat, and declared he was choaked: you may be sure that he eat no more.   The disagreeable sensation went off as suddenly as it came.   Perhaps I should have thought no more of this, as we continued to eat them without perceiving any thing in the least degree offensive; but, soon after, one of the maids was affected in the same manner. Are not these two instances sufficient to prove that something deleterious, though in a much lower degree, resides in this species, as well as in his brethren? And does it not confirm Linnæus's distribution of the genus *Solanum?* The seed-vessel seems to be the noxious part, and certainly affects some constitutions . . . .

Mr. Whitaker, in his account of Hannibal's passage over the Alps, says, that—'Potatos were known in Spain *before* Sir Walter Raleigh's time!' As the late Dr. Taylor used someimes to say,—'the fellow! why would he go out of his way to blunder?'—He means the *Convolvulus Batatas,* or Tuberous rooted Convolvulus.[a]

G. Ashby.

[a] [The Potato (*Solanum tuberosum*) now so common, was a curiosity in Botanic Gardens at the end of the sixteenth century, 1597. Much confusion has arisen from writers, unacquainted with Botany, not having distinguished between three plants to which the name of *Potato* was formerly attached, but which belong to different genera! The Battata, Potadee, or Potato, (*Convolvulus Battatas,*) a native of the East and West Indies, was common in Spain in the sixteenth century;—the Tuberous-rooted Night-shade, or Common Potato, (*Solanum tuberosum,*) a native of Virginia, introduced into culture by Sir Walter Raleigh, was a Botanical curiosity in 1597, as appears from Gerard's Herbal;—and the Canadian Potato of Parkinson, (*Helianthus tuberosus,*) a native of Brazil, absurdly called the Jerusalem Artichoke, seems to have been almost as common about 1629, as the common Potato is now.]

This anecdote may perhaps illustrate an extravagant opinion of Haller, (quoted by Professor Martyn in his MS. tour in Switzerland,) on the cause of the prevalence of goitres in the vallies of the Alps.—" Haller, the illustrious Haller," observes Professor Martyn, " after giving up all other theories, fixed upon *Potatos* as the cause of goitres!   This good Botanist had a systematic prejudice against this root, because it is of the same natural genus (*Solanum*) with the *Nightshades;* and the Potatos of the Valais are of a bad quality, having an acrid taste."—It seems probable that Haller had been led to the adoption of this singular theory, from having perhaps observed some instances of glandular affection of the throat, occasioned by the *root* of this bad variety of the Potato, similar to the effect recorded by Mr. Ashby, which followed eating the *seed-vessel.*

The publication of Dr. Smith's most interesting work, the " English Botany," which had been commenced in 1790, at this time much engaged the attention of British Botanists.   Respecting this admirable National Flora, Dr. Pulteney observes, (—in a letter to Professor Martyn, dated Feb. 4, 1795—) " I take the ' English Botany' and the ' Flora Rustica ;' the latter I *may* see an end to, but of the former I can scarcely expect it.   Have you fixed any limits, as yet, to the ' Flora Rustica ?'   I have often wished that the ' English Botany' had been in the hands of three persons ;—one wholly for the *Cryptogamia ;*—another, as is your plan, for *Agricultural* plants, extending it to the *Medicinal* ones ;—and the third for the remainder.   By this means a man might hope to have lived to see the matter completed.   This you may justly urge is dictated by selfish motives."—" Your partition of English Plants," (—replies Professor Martyn, March 2, 1795,—) " is a good one, and I wish it were adopted. *Cryptogamia* should form a work apart.   The *Medicinal* plants have been given by Woodville.   Our *Agricultural* ones will be finished in another volume or two ; but I wish to give more Grasses, they never having been figured together in the 8vo. size."—Professor Martyn's " Flora Rustica" was dropped, however, for want of support, in 1795.   The " English Botany," after twenty-four years labour, was completed in 1814, and exhibits 2592 figures of British Plants :—The publication of a Supplement commenced in August, 1829, which will probably add about 150 plants to this admirable national Flora.

About this time (1795) he wrote a short *" Description of the Hæmanthus multiflorus,"* or *Blood Flower,* (pp. 4, 8vo.,) accompanied by a folio coloured plate, by Nodder.   " This *Hæmanthus,*" he observes, " flowered with Mr. Parker of Fleet Street, at South Lambeth, in August, 1794 : this very handsome plant is a native of Sierra Leone, and is now recovered to the European stoves, after having been lost nearly two centuries."

In June, 1795, the publication of the " Gardener's and Botanist's Dictionary" commenced, with a number containing four sheets, and a part containing forty.

From Dr. Pulteney to Professor Martyn.

Blandford, Feb. 22, 1796.

Dear Sir,

. . . . I take in the " Dictionary," and I can say, with truth, that I stand *amazed* at the magnitude of your work, and the diligence and accuracy you have exhibited. It is much too good a book for those who are to enjoy the fruit of your labours—I mean, the booksellers; as to the public, they will stand indebted to you for a very long portion of time. I more and more lament that it could not have been thrown into a *systematic* form, so as to have been a regular Botanical work. I assure you I am not able to point out either error or improvement; but I hope, at the end, to see a list of all the genera disposed in *system*, for the use of those who may wish to study Botany regularly by it, which this addition will much facilitate. . . . . I was sorry to find your *Flora Rustica* dropped so soon. I hoped it might have been continued till all the Grasses had been completed. . . . .

R. PULTENEY.

In the midsummer term of 1796, the Professor read his *last* course of lectures at Cambridge; having continued them for a period of thirty-four years, without interruption, except in the years 1779, 1780, when he was abroad, and the year 1785, when his lecture-room was re-building. His state of health now unfitted him, in a great measure, for this annual labour; and, in truth, there was so little zeal for the study in the University, that it was scarcely possible to form a Class! Very few persons would have persevered *so* long in the effort to excite a taste for the elegant science to which his life had been devoted, but which, at *that* time, met with little favour at Cambridge.

From Professor Martyn to Dr. Pulteney.

Frith Street, Soho, Aug. 1, 1796.

Dear Sir,

I have continued in town to this time sorely against my will. To-morrow I set off for my country residence, at Pertenhall, in Bedfordshire, where I shall probably stay till Christmas. . . . .

You perceive that we get on reasonably well with the *Dictionary.* We are printing deep in H. . . I propose, among many other matters, in the Introduction, to have a complete *systematic* arrangement, as you suggest. . . . . The *Flora Rustica* may perhaps again be resumed, if peace, and times more favourable to science should ever return! . . . .

THO. MARTYN.

The '*Flora Rustica*' was not, however, resumed. An extended republication of this valuable work, adapted to the modern improvements and discoveries, both in *Agriculture* and *Botany*, seems well to deserve the attention of some scientific writer.

With the above letter concluded the correspondence

which had been maintained for thirty-six years between these two amiable men.   Dr. Pulteney's increasing infirmities rendered him incapable of further
literary exertion, and he died five years afterwards, at
the advanced age of nearly seventy-two.

In the summer of 1797, Professor Martyn communicated a second paper to the Linnæan Society; viz.:
—" *Observations on the flowering of certain plants. By
the Rev.* THOMAS MARTYN, *B. D., F. R. S., V. P., L. S.,
Regius Professor of Botany in the University of Cambridge. Read July* 4, 1797 " (Printed in the Transactions of the Linnæan Society for 1798, Vol. IV.,
pp. 158—163.)—The plants which form the subjects
of this paper, are the *Anagallis arvensis*, Shepherd's
Weather-glass; *Œnothera biennis*, Evening Primrose;
and *Hibiscus trionum*, Flower-of-an-hour.   The observations on their flowering were made by Professor
Martyn during the autumn of 1796.—" *Anagallis
arvensis*," he observes, " is a delicate wild plant, that
has long since attracted notice, as indicating a moist
atmosphere by the closing of its flowers, and the contrary by their opening : hence its name among the
country people, of *Shepherd's* or *Poor Man's Weather-
glass*.—The *Œnothera biennis* has been generally
regarded, from an early period, for the regular opening of its flowers in the evening; and has thence
obtained the name of *Evening* or *Nightly Primrose*.
While each flower is preparing for expansion, the
peduncle gradually diverges from the stem, and,
before the flower opens, arches downwards like a

swan's neck: the corolla swells out at bottom, and is
very apparently then between the leafits of the calyx,
which keep it close together for a considerable time
at top, by means of the hooks at the extremity of the
calycine leafits; till at length the corolla bursts its
bonds instantaneously, opens to a certain point, and
then, (having made a stand for a few seconds,) ex-
pands very slowly to its full extent. This critical
moment is very interesting to the botanical observer,
and may be seen with ease and pleasure, between six
and seven within the house, by gathering the flowers
and setting them in water. In hot weather the
flowers grow flaccid, and wither before noon the next
day; but in cool and cloudy weather they will last
two days.—The flowers of *Hibiscus trionum*, when
once expanded, continue often till they close finally;
they then droop. In warm weather the corolla folds
up wholly at night, and decays on the second day;
but in cool weather it will last a day or two half
folded up, but never opening so as to show the rich
purple eye at its base."—To these and many other
observations on this curious subject, are subjoined
tables of the heights of the Barometer, and the de-
grees of the Thermometer, from August 16th to
October 1st., with observations on the opening and
closing of the flowers during that period. From
these experiments it appears—that they do not open
at all, if the Thermometer be below 55° of Fahren-
heit,—nor *readily* if it be below 60°.

A very material alteration in the habits of his life took place the following year, by his removal from the busy sphere of the metropolis, to the quiet retirement of the lovely country village in which he passed the remainder of his days.   This change of residence was determined upon, in compliance with the wishes of his cousin, the Rev. John King, Rector of Pertenhall, Bedfordshire ; who, having recently lost his amiable wife, (mentioned above, pp. 50,—,) was desirous, in some measure, to supply the bereavement, by the permanent society of Professor and Mrs. Martyn at his parsonage.   Accordingly, in the spring of 1798, he removed to Pertenhall Rectory, which he rebuilt the same year.   In 1800, Mr. King* resigned that living, and presented the Rev. John King Martyn; and in 1804, he presented the Professor himself on his son's resignation.   On accepting that benefice, he vacated the Vicarage of Little Marlow.

From Professor Martyn to the Rev. G. Ashby.

Beaufort Buildings, Strand, Apr. 9, 1799.

My dear old Friend,

I shall see a good deal of Nichols, as we are likely to be joint editors of *Reliquiæ Stillingfleetianæ,*

* The Rev. John King died Oct. 6, 1812; having been Rector of Pertenhall forty-eight years, and resident there nearly sixty years. He was buried in the chancel, under the communion table. See some account of his family above, p. 50.

—not the Bishop, but the Gramineous Naturalist. . . I do not wonder that land-*owners* should be for the abolition of tythes; but that land-*holders* should be for it would be surprising, if farmers were not born to be duped. No sooner have landlords purchased the tythes, than squire Fox says to farmer Goose— " Farmer, I have been at great expense in ridding you of that abominable imposition—tythes; so you can now improve away, and reap all the advantage yourself; but you can have no objection to an increase of rent."—" Why master," replies farmer Goose, " to be sure our parson used to take three shillings an acre of us, and I had as soon pay it to your honour as to him."—" Aye, but," says squire Fox, " if *Parson* did not know what his tythes were worth, *I* do; and you must pay me an advance of ten shillings an acre in your rent, or quit."—Goose scratches his head, and remonstrates;—but Fox insists upon his pay or quit;—and so it must be! . . . .

THO. MARTYN.

From the same to the same.

Beaufort Buildings, May 10, 1799.

Dear Sir,

I thank you for your remarks on Stillingfleet. Nichols and I are going to dine at Greenwich, with Lieutenant Governor Locker, Stillingfleet's nephew, to settle preliminaries. Coxe has found some letters and papers; and we think that altogether we shall make up two volumes in octavo, of the same size

with the former.   I may perhaps, with your consent,
insert your remarks somewhere. . . .

THO. MARTYN.

The design of republishing Stillingfleet's Remains,
was dropped; but, a few years after, he gave very
material assistance to Mr. Coxe, who became the
editor of the works of that eminent naturalist.   (See
under the years 1809-11.)

The quiet retirement of his residence at Pertenhall,
afforded him the leisure absolutely requisite for finishing the great work in which he had been engaged
ever since 1785.   It was not, however, brought to a
conclusion without nine years *more* labour, from the
time of his coming to Pertenhall.   At length it was
completed, in the year 1807, and received the following title.—" *The Gardeners' and Botanists' Dictionary;*
&c. &c.   By the late PHILIP MILLER, *F.R.S., &c. &c.*
—*To which are now first added, a complete enumeration*
and description of all plants hitherto known, with their
Generic and Specific characters, places of growth, times
of flowering, and uses both medicinal and economical.
The whole corrected and newly arranged, with the ad-
dition of all the modern improvements in landscape-
gardening, and in the culture of Trees, Plants, and
Fruits, particularly in the various kinds of hot-houses,
and forcing-frames: with plates, explanatory both of
them, and the principles of Botany.   By THOMAS MAR-
TYN, *B.D., F.R.S., Regius Professor of Botany in the*
University of Cambridge.   London, 1807."   In " two

volumes," each of two parts; or, more properly, in *four* volumes, folio. The work is not paged.

The original work of Miller*, was in two volumes 8vo,, 1724; and was merely the germ of the succeeding eight editions, in folio. The folio editions are dated, in *one* volume, 1731, 1733, 1737, (a supplemental volume, 1739,) 1743, (the supplement not being reprinted;) in two volumes, 1748, 1752, 1759, 1768.—Between 1755 and 1760, he published 300 coloured plates to his Dictionary, in two volumes, folio. He made very great improvements in the successive editions of his work; and he states, that the number of plants cultivated at the time of publication, of the impression of 1768, was more than double those which were known in 1731. Miller's work was drawn up in the method of Tournefort. He was somewhat reluctant in adopting the system of Linnæus; but he followed it, in a great measure, in the edition of 1759, and introduced its nomenclature entirely in that of 1768.—He also published an Abridgement of his work, in two volumes 8vo., 1735; three

---

* PHILIP MILLER, F. R. S., was born in 1691. His father was gardener to the Company of Apothecaries at Chelsea, an office to which he himself succeeded in 1722. He died at Chelsea, in 1771; and a handsome pillar to his memory was placed in Chelsea Church yard, in 1815. His celebrated friend Houstoun, in 1730, dedicated to him a new Genus of plants discovered at Vera Cruz, and gave it the name MILLERIA; of which two species were figured by Professor John Martyn, in his "Historia plantarum rariorum,"—viz *M. quinqueflora*, (t. 41.), and *M. bifolia*, (t. 47. fig. 1.) See some account of his son Charles above, (p. 114.)

volumes 8vo., 1741, 1748, 1754; and one volume
4to., 1763, 1771.—This work undoubtedly laid the
foundation of all the *horticultural* taste and knowledge
in Europe; and Linnæus justly predicted that it
would become " a Lexicon, not of *Gardeners*, but of
Botanists!" It was held in such estimation on the
Continent, that it was translated into Dutch, (in 1746,)
into German, (1750-1,) and into French. "The latter,"
observes Professor Martyn, " has a fancy portrait of
the author in front, *in a bag wig and ruffles;* a cos-
tume which must appear truly ridiculous to such as
remember the *plain old fashioned English dress* in
which Mr. Miller always appeared!" Martyn's Pref.,
p. vii.

This greatly improved and prodigiously extended
edition of Miller's Dictionary by Professor Martyn,
ought, in point of fact, to have formed a new and
independent work ;—but the *name* of Miller was of
too much importance to the booksellers to be
omitted ; and the necessity of attending to their wishes
greatly cramped the Editor in many other respects,
particularly as regards the *arrangement*, which would
have been *systematic*, had he been allowed to follow
his own inclinations. Though he was compelled to
abandon the scientific classification, yet he bestowed
great labour on the arrangement of matter, which
had been much confused in all Miller's editions.
The Gardening was kept entirely distinct from the
Botanical part, in each article. He also added very
largely indeed to the enumeration of plants, inserting

descriptions of every species known at the time; with
the exception of the Mosses, Algæ, and Fungi, of
which the generic characters are given, together with
such species as are used in food or the arts. References are made to Linnæus, Reichard, Schreber,
Jussieu, Tournefort, Gærtner, Ray, &c. The *natural*
families, as well as the *artificial* classes and orders,
are pointed out; and figures are referred to. The
descriptions of the plants form a complete history
of the species. The directions for propagation and
culture are chiefly Mr. Miller's; but a great deal of
valuable matter has been added from other sources,
besides all the improvements in husbandry and horticulture, from the year 1768, (when the last edition of
Miller was sent out,) up to the date of publication.
The Philosophical articles were entirely new written.
In fact, the derivations of names, almost all the references, the generic characters, and most of the
descriptions, were *new*—that is, "not to be found at
all, or in a very different form, in Mr Miller's editions;" and whatever of any importance was added
by the Editor, was included between hooks, that the
reader might judge for himself of the actual value of
the new matter. "I am sensible," says the modest
Author, "that many will say, perhaps with a sneer,
that the work is nothing but a mere *compilation.*
The [strictly] *original* matter, indeed, bears so small
a proportion to the whole, that the Editor is content
it should pass under that humble title: but, if the
Gardener's and Botanists' Dictionary were to be con-

sidered as nothing more than a *digest* of what was known on Gardening and Botany at the end of the 18th century; or as a mass of information *collected* from the most scarce and valuable books, not accessible to the generality of readers, and written in languages little understood by practical men;—the Editor flatters himself that it may meet with the indulgence, if not the approbation, of a candid public. He is conscious, at least to himself, that, in the unwearied application of what talents he possesses, and the whole of that time which he could spare from the duties of his profession, during the last 20 years[*], to this laborious work, no attention or industry has been wanting on his part; and that he has strained every nerve to render it as complete in its kind as the nature of so extensive an undertaking will allow." Pref., pp. vi. vii.

After the completion of this laborious work, the Professor ceased to devote any considerable portion of his time to his favourite science; although his taste and enjoyment of it remained to the close of his life. The *only* undertaking, indeed, of this kind, which implied any serious labour, was the assistance he afforded the Venerable Archdeacon Coxe, in 1808 and 1809, in his three volumes of the "Miscellaneous

---

[*] [The actual period was 22 years; but during the first two years, the Professor did not devote so much time to the work as he did afterwards. He computed that, during these 20 years of hard literary labour, he had written "*as many thousand sheets of paper*."]

Tracts of Benjamin Stillingfleet." Professor Martyn
had himself intended to have republished these valuable pieces, so far back as the period of his residence
in Cambridge; and he had then made no inconsiderable collections with that view; but other engagements
postponed the design. In 1799, he resumed the
idea of becoming Stillingfleet's editor, (see above p.
221,) at the instigation of Captain Locker, the nephew of that learned Naturalist; but he again relinquished the plan, which was never resumed.   When
Mr. Coxe, at length, turned his attention to the
same subject, he naturally applied to his old acquaintance, Professor Martyn, for assistance in that part of
the work in which he avowed his own deficiency of
information.   The Professor, though at that time in
ill health, and unable to devote more than three hours
in the day to business of *any* kind, yielded to Mr.
Coxe's solicitations with his usual urbanity and frank-
ness.   He not only assisted the editor, throughout the
whole of the work, by various communications, by
looking over the MS., and by transcribing many MS.
remarks* made by Stillingfleet himself, and which were

---

* These were contained "in six thin quarto paper books, in which
Stillingfleet entered extracts from works on agriculture, &c., that
he was reading, and sometimes *observations* of his own," made
about 1764.   The latter, Professor Martyn classified; and he
transcribed them, so arranged, for Mr. Coxe's use.   He also sent
Mr. Coxe a paper on the irritability of flowers, translated by Stillingfleet, from the Italian of Count Giovanni del Colvola, which
had never been published.

in his own possession; but he drew up some original papers, which are printed in Mr. Coxe's volumes,— viz. 1. " *Additional Observations on Grasses,*" supplementary to those of Stillingfleet (Vol. II., Part II., pp. 305—357); and 2. " *Observations on the times of leafing, flowering, and the fall of the leaf, of certain plants and trees, from the year* 1775 *to* 1809, *arranged in tables,*" (Vol. II. 493—504), intended as an Appendix to Stillingfleet's " English Calendar of Flora." This last paper is particularly interesting, and the subject seems deserving of more extended investigation. " My Table of the leafing of trees, and the fall of the leaf," writes the Professor to Mr. Coxe, (Feb. 10, 1808), " will be *imperfect;* but every set of observations, provided they be tolerably accurate, contributes *something* towards drawing practical conclusions." Professor Martyn's notes in these volumes are marked with the letter M. Mr. Coxe gratefully acknowledged, in his preface, the assistance he had received, (see Vol. I., p. 11.) The following short extracts from his correspondence with Archdeacon Coxe at this time, will show that he had not discarded Botany from his mind, though he had taken leave of it as a serious pursuit :—

(Feb. 25, 1808). " Gesner's are mere hints, and it is not probable that Cæsalpinus had seen them. The latter is clearly the author of method or arrangement. No man had any idea of it before him;

I mean, as applied to the *vegetable* kingdom. Cæsalpinus's arrangement is founded wholly on the seed and seed-vessel. It ought to be remarked, in justice to this eminent logician and Botanist, that his system contains many of the finest *natural* Classes; as, the *Grasses*, in the 4th Book; the *Umbellated* plants, in the 7th; the *Bulbs*, in the 10th; *Asperifoliæ*, and *Verticillatæ*, in the two sections of the 11th; *Compositæ*, in the 12th, 13th, and 14th; and *Cryptogamia* in the last. I do not mean that you should insert all this in your Sketch." . . . .

" Whatever other merits Morison's great work might have, as a system or arrangement it was worth nothing, and was never followed. Whereas Ray's was the best arrangement till Linnæus arose, and it was followed, and improved, by Hermann, Knauth, Volckamer, Commelin, Boerhaave, Dillenius, and particularly by my father. Tourneforte's system, however specious and elegant, will seldom enable a Botanist to *investigate* a plant. Foreigners have done great injustice to Ray. Even Linnæus and Pulteney have not done him *complete* justice. His distinctions are clear and logical; and he has preserved as many natural classes entire, as Linnæus himself. I could easily convince you of this; but it would take up too much room, and not answer your purpose." . . . .

(March 26, 1808.) " The Egyptian Sycamine of Theophrastus, (Hist. Plants, Chap. I.) is the *Ficus*

Sycomorus[a], or Sycomore of Scripture; a huge tree, common in lower Egypt. The coffins in which the mummies are preserved, were made of this wood.[b]"—

(April 27, 1808.) "You see it is not easy to procure effectual assistance, where it might be had most easily. *I* solicited aid, both in print, and by letter; but obtained none worth speaking of;—so I gave it up, and worked alone, like a dray-horse, for twenty years! If you print Stillingfleet's Flora's Calendar, from Theophrastus, it might be well to remark, that Μελια is *Fraxinus Ornus*, or, Manna Ash,—not *F. excelsior*, or Common Ash, which is not found in Greece, so far as we know; and that the Fir called Πευκη, seems probably to be the *Pinus maritima*." . . . .

(May 16, 1809.) "Almost the whole merit of Klijog's[c] husbandry appeared to me to consist in the

[a] [This is the *Mulberry-leaved Fig*, or the *true Sycomore ;*— the tree usually called the *Sycomore* being the *Acer Pseudo-platanus*, a species of *Maple*. The *Ficus Sycomorus* is of great use, in a scorching climate, to shelter travellers in the deserts. The wood is reddish, light, and does not decay for ages. It is figured in Plukenet, Phyt. t. 178. f. 3.]

[b] [Some late writers have affirmed, that the mummy coffins are made of the wood of the *Evergreen Cypress*, (*Cypressus sempervirens*.) On what authority this assertion, or that of Professor Martyn, rests, does not appear.]

[c] [This famous Swiss peasant resided two leagues from Zurich. Professor Martyn visited him in June, 1779. (Martyn's MS. Journal, Vol. I., p. 270.) Dr. Hirzel's account of him, (4th edit.) was printed at Lausanne, in 1777.]

judicious mixture of different soils; and I had a long
conversation with him on the subject. You remember that he is celebrated by Dr. Hirzel of Zurich,
under the name of '*Socrate rustique.*' Certainly much
improvement might be accomplished this way, if soils
of opposite qualities were within reach of carriage;
which is seldom the case. . . . .

" Our English word, *grass*, was anciently of very
extensive signification. To prove this, I need only
refer you to our English translation of the Bible.—
' If God so clothe the *grass*[a] of the field, which to-
day is, and to-morrow is cast into the oven;' (Matt. vi.
30;) alluding to heating their ovens[b] with the stalks of
herbaceous plants.—' As the flower of the *grass* he
shall pass away;' (James i. 10)—consider the eleventh
verse." . . .

[a] Χορτος. Wetstein remarks that, by this word the Greeks denote
" grass, corn, and *flowers.*" Schleusner translates it, in this place,
"*flos* agrestris;" and Parkhurst observes, that "it is certainly
designed to include the *lilies* of the field, of which our Lord had
just been speaking." Sir J. E. Smith supposes that this " *Lily* of
the field," is the *Amaryllis lutea,* or Yellow Amaryllis, which
grows abundantly wild in the fields of the Levant. A writer in the
Christian Remembrancer for 1819, (Vol. I., pp. 73—77,) contro-
verts this opinion, (but not by any forcible reasons,) and endeavours to show, that the *Lilium candidum,* or *White Lily,* is the
plant in question; others have supposed it to be the *Fritillaria
imperialis,* or *Crown Imperial :*—it is very doubtful whether the
last two plants belong to the Flora of Palestine.

[b] [See Harmer's Observations, Vol. I., p. 264.]

(September 17, 1809.) " In compliance with your request, and that of Mr. Stackhouse, I send you the substance of what my father has said on the *Cytisus* of Virgil." [Professor John Martyn's note on Georg. II. 431., is here quoted at length, in which he endeavours to show, that Virgil's *Cytisus* is the *Cytisus Maranthœ,* as is the *Cytisus* of Theophrastus.] " You will observe, that *Cytisus Maranthœ* is not a *Cytisus* of Linnæus, but his *Medicago arborea,* or Moon Trefoil, a shrub of the same genus with Lucerne; and I little doubt of this shrub being the *Cytisus* of Virgil. I figured and described it in p. 100, of my ' Flora Rustica.' . . . It is *more* probable that the *Cytisus* which Theophrastus mentions, as approaching to Ebony in hardness and colour, *may be* the *Cytisus Laburnum,* as Mr. Stackhouse thinks. It is Honorius Bellus who relates (what is mentioned, without specifying the authority, by my father) that the Turks make the handles of their sabres, and the Caloyers their beads, with the wood of *Cytisus Maranthœ,* i. e. of *Medicago arborea :* so that, after all, *this* may be the *Cytisus* in all the three places of Theophrastus, quoted in the note above. If one were but in the Isle of Rhodes, it might be ascertained."—

In 1812-3 it was proposed, by the members of the Linnæan and Horticultural Societies, to erect a monument to the memory of the celebrated PHILIP MILLER. (See above, p. 224, note.) Professor Martyn, having been applied to on this subject by Sir Thomas Gery

Cullum*, Bart., addressed the following letter to his old and valued friend :—

Pertenhall, Jan. 25, 1813.

My dear Sir,

I shall most willingly contribute, what you or our other friends think proper, to set up a memorial for Mr. Miller. He well deserves it, both from the Linnæan and Horticultural Societies. *I* ought to subscribe, not only as a member of the first, but for the kindness he showed me when a boy and a youth, and for the good salads, and Cantelupe melons, and venison, with which he has treated me. I believe nobody else now remains who can give this reason for subscribing. You remember that Charles Miller, soon after his return from abroad, expressed an intention of putting up a monument for his father, and desired me to draw up an epitaph, which I did. I think you asked to see it, but I have not kept any copy, and I have now so far forgot it as not to recollect whether it was in Latin or English. I would not now, in the weak state of my head, undertake to write one ; it is a species of composition so nice and delicate, must

---

* His acquaintance with this amiable Botanist, (who still survices at an advanced age,) began very early. " I had the honor," writes Sir T. G. Cullum, " of receiving a note from him in the spring of 1763, politely inviting me to his Botanical Lectures, and from that time, a mutual friendship and occasional correspondence, subsisted between us."—Letter from Sir T. G. Cullum, Bart., to the author of this memoir, Feb. 16, 1829.

have such elegant turns and be so classical, that it is dangerous to meddle with it.   When I lived in London, I was sometimes requested to draw up an *advertisement :* to which I replied, that I had rather write a *book :* so I say of an *epitaph*, if one is to launch out beyond the mere chronicle. . . . . We united in repairing Mr. Ray's monument.   Let us then unite in another good work, for an honest man who left no memorial of himself but his useful writings.   It is a pleasure to record the memories of men who have been useful.   He was certainly *unique* in his time. . . . Neither you nor your correspondents say where the tablet is to be erected.   The argument against Chelsea Church gains strength as that fabric weakens.   I remember, when I was a boy, it was voted to be ruinous, and would then have been pulled down, if my father had not buttressed it up with all his interest.

Flora Græca gets on so slowly, that I have little chance of seeing much more of it.   When are we to have the *British Cryps ?* [a]   I am not much interested in them, though I do not go so far as Davies as to say they are all alike. . . . .

We have been full of affliction.   On the 27th of July died my son's young, amiable wife.   On the 6th of October, our old friend the Rev. John King [b], with

---

[a] [Alluding, probably, to the Volume of Dr. Smith's *Flora Britannica*, in which the *Cryptogamia* were expected.]

[b] [See above, pp. 50, 180, 220, 221.]

whom I have lived in friendship ever since I can remember any thing, departed from his post here, which
he had rigidly kept for sixty years. . . . .

Tho. Martyn.

About this time[a] he volunteered his services to Mr.
Bray, by contributing " *A List of rare plants in
Surrey,*" for his enlarged edition of Manning's History
of that County.　This List is printed in Vol. III., pp.
lxv.—lxx. of that work, published in 1814.　The
plants noticed, are chiefly those he met with in the
years 1758-9, when visiting his father at "the Hill
House," Streatham.

The following extract from a Letter[b] to the Rev.
Weedon Butler, shows the peaceful state of his
benevolent mind at this period.　It was occasioned by
an application to him for his acceptance of Mr.
Butler's son as the Minister of Charlotte Chapel,

---

[a] In kind compliance with the wishes of a friend, he gave his assistance in the publication of the following volume :—" A Treatise
on Human Happiness, by the late Rev. W. Stevens, D. D.; Edited
by the Rev. Thomas Martyn, B. D., F. R., & L. SS., Regius Professor of Botany in the University of Cambridge."　London, 1813.
8vo. pp. 276.—He was, however, (as he observes in an Advertisement,) responsible for nothing in the work itself, with the exception of four pages of *Notes* at the end.

[b] Printed in Nichols's Lit. Anec., Vol V., p. 752.　Mr. Butler
was proprietor of one-quarter part of Charlotte Chapel.　He had
been a pupil of the unfortunate Dr. Dodd, of whom he had purchased his share in that Chapel.　He died in 1823.

Pimlico, of which Professor Martyn was then the principal proprietor :—

From Professor Martyn to the Rev. Weedon Butler, Sen.

Pertenhall, 28 Sept. 1814.

My dear Sir,

    I ought to have answered your favour of the 8th instant sooner, and might certainly have done it, but time runs on insensibly, and my ability for writing is very small.   As I enter on my 80th year on Tuesday next, I have reason to be thankful that I am able to read or write at all, that I can walk about my premises, and drive myself in my gig ; and, above all, that I can yet preach every Sunday.   I was truly gratified to find that you intend removing to Gayton; both because the retirement to so pleasant and healthy a situation, and quitting the bustle and fatigues in which you have been engaged, must be very agreeable at your time of life; and, also, because the flock will not be left to a common hireling, but will, I am well persuaded, be duly fed with the most salutary food. This is an object which must be near the heart of every conscientious clergyman.   It is melancholy to see several of our neighbouring parishes without so much as a resident Curate, served irregularly once on the Sunday in haste. . . . . In this parish the Rectors have been constantly resident ever since the Reformation.   For the last 120 years my family have been both Patrons and Rectors; and we have considerable influence in it. . . . .

THO. MARTYN.

In July, 1815, an afflictive event, which occurred in the Professor's family, brought the writer of these pages, for the first time, into personal acquaintance with him, as the temporary Curate of his parish. He cannot look back upon this period, without the most lively and interesting recollections of the piety, urbanity, and cheerfulness which were so delightfully manifested in the aged Rector of Pertenhall;—qualities which, existing in a mind so richly stored with scientific information, rendered him a charming companion, even to those who were far below him in years. Being now engaged in preparing this Memoir for the press, in the parsonage which then owned Professor Martyn as its inhabitant, he could almost persuade himself that he still hears the very voice which fourteen years ago welcomed his Curate to his abode, and that he still beholds the venerable figure of his departed friend?

One of the most pleasing features of Professor Martyn's mind, was his easy condescension to young persons, especially when he saw them desirous of gaining information from him. The following letter affords a good example, both of the simple, elegant manner in which he delighted to communicate instruction to his young acquaintance, and also of the vivacity of his intellect at this advanced period of his life—for he had entered on his *eighty-first* year the day on which it was written.

From Professor Martyn to Miss Christiana Gorham.

Pertenhall, Oct. 4, 1815.

My dear Madam,

I send you the seeds, which your brother requested, of *Atropa physaloides**, or *Peruvian Deadly-Nightshade.* They cannot come to him, I am persuaded, through any hands so agreeable as yours; and I will not disguise that I am happy in taking an opportunity of testifying my esteem for you, and the pleasure I took in your conversation. . . . . The seeds might have been inclosed in a nut-shell; but, as you seem curious in such matters, I was desirous you should observe the apparatus which nature has provided for the protection of the seed; there is first a shell, which, being thin and brittle, is, next, itself protected by the permanent calyx, in form of a regular pentagon, with five bastions.—I presume you do not send the seeds into Scotland, for they will not vegetate there, except in a hot-bed: indeed, even here, in untoward seasons, they scarcely flower soon enough for the seeds to ripen; it is safer, therefore, to commit a few of them to a gentle hot-bed. You are aware that the plant is an annual.

* [The *Atropa physaloides* of Linnæus and Jácquin, is the *Nicandra physaloides* of Jussieu; a native of Peru and Chili.]

On the top of the little case, I have put into a paper a few capsules of the *Flax*[a];—not for their rarity, but because they are a comment upon Exodus ix. 31.[b] You will observe that they have the shape of a *bowl*, and they were called *bowlles* by the writers of Queen Elizabeth's days. If you have not observed the Flax, you may sow it in the spring, and be pleased with the elegance of the plant, and the beauty of the flower. It is the most interesting of plants,—for it constitutes one of the principal ornaments of your sex—is found

[a] [*Linum usitatissimum*, Common Flax.]

[b] [" Y⁰ Flex and yᵉ Barli was hirt, for yᵉ Barli was greene and yᵉ Flex hadde buriollned yanne *knoppis*." Wicliffe's Bible, about 1390, MSS. Brit. Mus., King's Lib. I. C. VIII.—" The Flax and Barlye were smytten; for the Barlye was shot up, and the Flax was *boulled*." Coverdale's Bible, 1535; and the Bishop's Bible, 1568.—" The Flax and the Barley was smitten; for the Barley was in the ear, and the Flax was *bolled*." King James's Bible, 1611.—These terms may be well illustrated by a passage from a Botanical work of Queen Elizabeth's time. " FLAX. After that the flowers be gone, ther come furth round *knoppes*, or *heades*, called in Northumberland *bowles*." Turner's Herbal, 1568.]

[c] [" This valuable plant," (observes Professor Martyn, in his edition of Miller's Dictionary,) " is supposed to have been derived originally from those parts of Egypt which are exposed to the inundations of the Nile. In the earliest record we have, (Exod. ix. 31) *Flax* is mentioned as a plant cultivated in that country; for which reason antiquarians have been surprised to find the vestments of mummies made of *Cotton*. It is highly probably, however, that mankind made thread of *Cotton*, before the use of *Flax* was discovered; the former being produced in a state ready for spinning; whereas the latter requires a long process before it can be brought to that state."]

always at bed and at board, and is the chief instru-
ment of cleanliness to both sexes. . . . .

Do me the credit to believe, my dear Madam, that
I am, with true regard.

Your faithful humble Servant,

THO. MARTYN.

The following letter was addressed to Sir Joseph
Banks, (whose friendship he had for many years cultivated,) on transmitting him some Botanical correspondence which he thought worth perpetuating.
These papers consist of a great number of letters
from Dr. Patrick Blair, a few from Mr. Wilmer, one
from Mr. Collinson, on the *Limodorum altum*, one
from Mr. Smith, of Croydon, on the discovery of
Scrophularia vernalis, and a few from Mr. Gilkes,—
to Professor John Martyn; together with a great
many from Dr. Pulteney, a few from Philip and Charles
Miller, Mr. G. Ashby, Sir T. G. Cullum, Bart., Mr.
T. J. Woodward, and Mr. Relhan, one from Dr. Darwin, one from Dr. John Hope, and one from Mr. Pennant,—to Professor Thomas Martyn.

From Professor Martyn to Sir Joseph Banks, Bart.

Pertenhall, 14 March, 1817.

Dear Sir,

Herewith I send you a packet of old letters, some
of which seem worth preserving; but you are at full
liberty to dispose of them as you please. I am sorry

R

to hear that you have been a great sufferer from illness.   Last year I thought that I was moving off this stage ; but this year I have rallied again, and can enjoy my books at home, for I never go out.

I see that Mr. Rennell is coming out with a work, which I hope to read with as much pleasure and profit as I did that on Herodotus. Xenophon, I must confess, is my favourite among the Greeks.   When you see Mr. Rennell, will you ask him if he has ever particularly considered the use of chariots in war—whether there be any satisfactory treatise on the subject—when they were completely superseded by cavalry—and whether there be any ground for supposing, from the use of them here in Julius Cæsar's time, that the inhabitants of our Southern Coast were a colony from Asia ?   One cannot help comparing Cæsar's brief account with Homer.

THO. MARTYN.

Since the year 1796, Professor Martyn's increasing infirmities had prevented him from continuing his annual course of lectures[a].   His beloved science had, indeed, been so little patronised at Cambridge[b], that a

---

[a] See above, p. 217.

[b] This reproach no longer attaches to the University. Among other improvements at Cambridge, it is pleasing to remark the encouragement which is *now* given to Botany. On the 27th of February, 1829, an important Grace passed the Senate, by which it is required that all candidates for the degree of Bachelor in Medicine shall attend one course, at least, of lectures in *Botany*, and be examined by the Professor in that science, before admission to the performance of their exercises in the Schools. An increasing *taste*

less zealous Botanist might have been excused, at his advanced age, from making any further effort on the subject; however, he did not allow discouragement to interfere with his sincere desire to awaken the attention of the University to his favourite study. The author of these pages has often conversed with the Professor on this topic; which he never touched without an animation, which showed how greatly his own mind had been interested and improved by the study of this most elegant department of Natural Philosophy. The language in which the illustrious Ray had expressed his feelings, on the neglect of Botany at Cambridge, in the middle of the seventeenth Century, exactly describes the sentiments of Professor Martyn, on the same subject, at the beginning of the nineteenth :—" I am ashamed and grieved whenever I reflect, that this choice department of *Natural Philosophy* and of *Natural History*,—so useful, I might almost say essential, to human life—should alone remain neglected and uncultivated, while all other sciences and liberal pursuits flourish and are cherished among us[a]!"—In the winter of 1804, he recommended Mr.

for this delightful study is also manifesting itself in the University. Professor Henslow's Botanical Class averages seventy students, and about thirty accompany him in his herbarizing excursions in the neighbourhood.

[a] " Puduit sane nos piguitque, quoties animadvertimus, potissimam hanc *Naturalis Philosophiæ et Historiæ* partem, communi vitæ adeo utilem et tantum non necessariam,—cum aliæ omnes scientiæ et ingenuæ disciplinæ apud nos florerent et illustrarentur, —solam jacere neglectam et incultam!" Raii Catal. Plant. circa Cantab. nasc. (Præfat.) Cantabrigiæ, 1660.

Relhan to attempt to form a Botanical class the ensuing spring. "A violent attack upon my lungs" (he
writes to Mr. Relhan, Dec. 22, 1804,) "during the
summer, has totally unfitted *me* for reading any more
lectures. If the Governors of the Garden approved,
I could have wished that it might be made worth
your while to read a course; but there does not seem
to be Botanical spirit enough in the University to
recompense you for the trouble that would attend it."
This wish came to nothing. In 1813, Sir J. E. Smith,
the President of the Linnæan Society, applied to
Professor Martyn for his sanction to make another
attempt to supply the deficiency of public lectures.
He was strongly urged to this undertaking by the late
Sir Joseph Banks. The Professor cheerfully complied
with this request, and voluntarily offered to resign the
Walkerian[b] Readership in Dr. Smith's favour. To
this proposal the Vice-Chancellor was favourable; but
the other trustees declined giving their consent, on
the very reasonable ground that they ought not to be
fettered in the choice of a successor; and on the fur-

---

[a] This accomplished Botanist died on the 17th of March, 1828,
in the sixty-ninth year of his age.—The SMITHIA *sensitiva*, an
East India plant of the Diadelphian Class, commemorates his name.

[b] Professor Martyn held three offices; he was—1. *Titular Professor of Botany*, appointed by a Grace of the Senate in 1762,
without salary;—2. *Dr. Walker's Reader in Botany*, appointed
by trustees, also in 1762, without salary, but with the use of the
Garden;—3. *Regius Professor of Botany*, appointed by Royal
patent in 1793, with a pension of £100, afterwards increased to
£200.

ther allegation (the liberality of which, in the absence
of competition of candidates, was much canvassed) that
Dr. Smith was not a proper object of their patronage,
not being a member either of the Church of England
or of the University*. Thus the matter rested till the
year 1818, when Professor Martyn,—with the simple
desire of providing Botanical instruction for the University, and guided by the precedent of lectures
having been read by his own father from 1728 to 1732,
to supply the negligence of Professor Bradley,—expressed his wishes that Sir J. E. Smith might be
allowed to lecture as his *deputy*. On this occasion he
wrote the following letter :—

From Professor Martyn to Sir J. E. Smith.

Pertenhall, March 14, 1818.
My dear Sir,

   The season approaches when I feel an annual
regret that, in consequence of my age and infirmities, I am unable to fulfil my duty as Dr. Walker's
Reader, in giving a course of Botanical lectures. If
you could, consistently with your other engagements,

---

* This was not without precedent. Bradley, the first titular
Professor of Botany, in 1724, and Rolf, the first *Professor of
Anatomy*, in 1707, were not members of the University. On the
other hand it was said, that these appointments were made on the
introduction of those sciences into Cambridge, when, perhaps, it
was difficult to find qualified persons among the Members of the
University.

undertake to read a course next term, I should
esteem it a great favour done to me personally, and
I have no doubt of its being well received by the
University. You are aware that you must have the
sanction of the Vice-Chancellor, who, I am persuaded,
will be ready to give the University an opportunity of
profiting by your instructions; as he doubtless knows
that you take the lead in the science of Botany in
this country, and that your reputation is too well
established to need any recommendation from *me*.
As far as my power extends, I am happy in giving you
full authority to take such specimens of plants and
flowers as you think requisite for your lectures;
together with the use of the lecture-room, at any
time or times that may be convenient: *always under
the controul of the Vice-Chancellor*, and with a complete reliance on your discretion in the use of the
Garden.

Sincerely wishing it may suit your convenience to
comply with this my request,

I am, &c.

THO. MARTYN.

The Vice-Chancellor, (Dr. Webb, Master of Clare
Hall,) having given his sanction to this proposal,
Dr. Smith issued a printed notice of his course, to
commence on Monday the 6th of April. However,
on Saturday the 4th of April a written remonstrance
was sent to the Vice-Chancellor, signed by eighteen
tutors of Colleges, stating their decided disapproba-

tion of their pupils attending the public lectures of
" any person who is neither a member of the University, nor of the Church of England." Although
the Vice-Chancellor himself, and several members
of the Senate, had declared their intention to attend
the lectures; so that Sir James Smith might still
have formed a class entirely independent of the
controul of the gentlemen who signed this protest ;—
yet he prudently determined to give up his plan,
rather than involve the University and himself in a
disagreeable and personal contest. It would have
been well had the matter ended by Dr. Smith quietly
retreating from a scene in which he had met with so
much (and, as not a few thought, such unmerited,)
opposition ; but, unhappily, the disappointed President of the Linnæan Society immediately commenced
a paper war, in the prosecution of which he indulged
in unbecoming attacks on private character, and introduced into his pamphlets some highly exception-
able sentiments of a Socinian complexion. With all
the circumstances of this lamentable controversy[a]

---

[a] This controversy consisted of four pamphlets ; 1. " Consider-
ations respecting Cambridge, more particularly relating to its
Botanical Professorship, by Sir J. E. Smith, M. D., F. R. S., &c.,
President of the Linnæan Society. London, 1818." 8vo. pp. 60.—
2. " A Vindication of the University of Cambridge from the re-
flections of Sir J. E. Smith. By Rev. J. H. Monk, B. D., Fellow
and Tutor of Trinity College, and Regius Professor of Greek in
the University of Cambridge. London, 1818." 8vo. pp. 95.—3. " A
Defence of the Church and Universities of England, against such

the writer of this memoir was fully conversant ; being
then resident on his Fellowship in the University, and
well acquainted with both Dr. Smith and Professor
Martyn ; but he declines pursuing so disagreeable
a subject further, nor would he have touched upon
it but for the purpose of showing that the Professor
himself was *altogether* unconnected with it *as a party.*
Dr. Smith might (as was alleged,) have interested
views of ultimately succeeding to the Botanical Professorship ; and his opponents might object to his
lecturing, lest he should obtain a popularity preju-
dicial to other candidates for that office, whenever it
might fall :—with *such* views, Professor Martyn did
not meddle ; his simple desire being to provide *
adequate instruction for the University in a science
in which he had exerted himself so many years, but
which his age and infirmities prevented him from
promoting any further by personal effort. It was,
however, painful to this venerable man to observe

injurious advocates as Professor Monk, and the Quarterly Review
for January 1819. By Sir J. E. Smith. London, 1819." 8vo.
pp. 107.—4. "Appendix to a vindication of the University of
Cambridge, &c. By J. H. Monk, B. D. Cambridge, 1819." 8vo.
pp. 55. *printed for private circulation ; not published.*

* In 1821, he made one more attempt of the same kind, by
acceding to the wishes of a gentleman to become his deputy, to
whom the University could have no reasonable objection, as he was
chosen the Botanical Professor on Mr. Martyn's death :—but Dr.
Walker's trustees declined giving their consent. In the winter of
the same year, he offered to resign the Walkerian Readership,
without having any individual in view as his successor.

an angry controversy raging, connected with preferment of which he could not *long* continue to disap-
point the most anxious expectant; and still more
painful to see his eminent Botanical friend forsaking
"*the flowery path*," in which he had been gathering such well-deserved garlands, and pursuing a
thorny way in which he reaped nothing but discredit
and disquietude! But although the ambitious and
selfish feelings, which were more or less called forth
during this unseemly strife, could not be contemplated
without some degree of emotion by his delicate and
liberal mind,—they did not disturb its habitual tranquillity. Referring to these matters, a distinguished
Botanical friend[a], who in early life had attended his
lectures, reminded him, by letter, " that he was too
much of a Philosopher to be keenly affected by them :"
—" if not too much of a *Philosopher*," he replied, (in
language elegantly allusive to their common studies,)
" I am too much of a *Christian* to be affected by any
thing that is passing at Cambridge, or any where else;
for, standing as I do on the very verge of both worlds,
I am looking forward to

'—yellow meads of [*] Asphodel,
' And Amaranthine bow'rs.'"[b]

The following letter, written to Dr. Smith, about
three years after this period, brings the late President

[a] Sir T. G. Cullum, Bart.   [b] Pope's Ode for St. Cæcilia's Day.

[*] — " They came into the meads
[d] Of *Asphodel*, by shadowy forms possess'd."
Homer's Odys. xxiv. 13. (*Cowper's Trans.*)

of the Linnæan Society before us in a more pleasing
character, and contains just remarks on one or two of
his later works.

From Professor Martyn to Sir J. E. Smith.

Pertenhall, March 9, 1821.

My dear Sir,

I received your book, which gave me great
pleasure. Your " *Grammar* " plainly speaks the hand
of a master; concise, yet full; remarkable for clear-
ness and neatness. Small as it is in size, it must
have cost you some time and attention. I smile,
sometimes, when I meet with the miserable, incorrect compilations and imitations of your former work
[" *Introduction to Physiological and Systematical Bo-*
tany."] When your intended *Flora* makes its appearance, the British Botanist will find every thing
that he wants in these three works of yours.

.I am not such a bigot as to think lightly of the
natural orders, imperfect as our present knowledge
of them is. Had I been younger, that very circumstance would have incited me to pursue so delectable
a subject, and I hope *you* will continue to do it. I
am only sometimes vexed when they would fain persuade me that the natural system may *supersede* the
artificial.

Some time since, I sent the Curator of the Cambridge Garden, a parcel of seeds which I had received,
with a box of specimens, from Van Dieman's land;
but I have never heard whether they succeeded.

Some of them grew with me under many disadvantages. . . .

The winter has been very kind. I have nothing very tender; but, in this cold soil, the *Bay* and *Aucuba Japonica* are cut almost every winter,—during *this*, they have been untouched.

It gives me much pleasure to have lived to see such vast improvements in Botany and Horticulture, as have been made in the last twenty years; though they have rendered *my* poor performances of no value. My health is still good; but I grow sensibly weaker in body, and memory and recollection are decaying. With all this, I am thankful that my eyes are yet efficient, that my spirits are good, and that I can still enjoy conversation with those who will condescend a little to my deafness. This is much for almost eighty-six.

I thank you for your information about Sir Thomas and Lady Cullum, of whom I have not heard for some time. With my hearty wishes for your health and happiness,

I remain, &c.

THO. MARTYN.

From the same to the same.

Pertenhall, April 6th, 1821.

Dear Sir,

I dispatched yesterday the box of specimens, from Van Dieman's land. Of many of the species

there is such abundance of specimens, that you may
oblige all your friends. They are of little use to *me,*
who am now attending to other matters. I wish the
box may arrive before you set out on your spring excursion. You are happy in having Lady Smith to
assist you in fastening your specimens; the ladies'
fingers are better adapted to all nice works, than
ours.

I am glad to hear that the *Flora Græca* is going on,
but shall not see the end of it. I take it; but how
many purchasers can there be of so expensive a work!
It is, however, very tempting from its elegance.

Biography is so much in request, that I should
think it would answer to reprint the lives of *Botanists;* especially if you would revise and complete
them. What has been said of *me,* in the European
Magazine, and the Biographical Dictionary of living
Authors, being incomplete and incorrect, I have been
lately drawing up Memoirs of myself, for the information of *my family;* not being vain enough to
think that any thing I have done can interest *the
public.* . . .

THO. MARTYN.

Of the magnificent work, the *Flora Græca,* mentioned in the preceding letter, he had thus expressed
himself, about ten years before:—" This classical
work may truly be denominated *fortunate,* except in
the premature death of the excellent Author; for no
man could have been selected more able to ascertain

the plants of Greece than *Sibthorpe;* or so fit to be the editor of them as *Smith;* no draughtsman could have been chosen like *Bauer;* no engraver like *Sowerby!*" (Stillingfleet's Misc. Tracts, edited by Coxe, 1811, Vol. II., p. 385. *Martyn's note.*)

From the same to the same.

Pertenhall, Nov. 16, 1821.

Dear Sir,

    I have been reading your *Selection of the Correspondence of Linnæus, &c.;* and have been, perhaps, more interested in the work than most others. For, though the chief part of the correspondence is antecedent to my Botanical life, yet most of the parties were familiar to me in my early days; though I had personal acquaintance with few of them. These letters show how dead Botany was, in England, in the middle of the last century; when Collinson and Ellis, two men not professionally scientific, but engaged in commerce, were Linnæus's principal English correspondents! His system can hardly be said to have been *publicly* known among us, till about the year 1762, when Hope taught it at Edinburgh, and I at Cambridge, and Hudson published his Flora.

    I was much amused with the sparring of two such great men as Linnæus and Haller, till it became serious on the part of the latter: but it discovered plainly the great superiority of mind and temper in the former.

What letters I had relating to Botany, I gave to Sir Joseph Banks; but there were few within the compass of those which you have edited, except Dr. Blair's, and they had no reference to Linnæus. I also gave Sir Joseph the documents relating to a little Botanical Society, early in the last century, of which my father was secretary. . . .

Go on, and prosper, my dear friend, in your *Flora Græca,* and your other most useful works, in promoting *the most delightful of all the sciences;* and believe me to be, &c.

THO. MARTYN.

These letters,—which were probably among the latest addressed to his scientific friends,—afford a pleasing evidence of the vigorous state of the Professor's mind, at the advanced age of 87. Indeed, his faculties seem to have been scarcely impaired at this period, nor would it have been easy to fix upon any other decisive indications of the infirmity of age, than the increasing feebleness of his constitution. Happy in the resources of his own mind, in his reading, and in the society of the different members of his family, all of whom were placed in the immediate vicinity of his quiet parsonage; his declining days passed on in a tranquil and even tenor; and the current of life ebbed away by such gentle gradations, that its retreat could only be perceived by a comparison of distant intervals. His bodily activity was such, that, till within a few months of his death, he was able to take air and exercise, and to make frequent visits to his beloved and aged sister, Mrs. Elizabeth Longmire, at the neighbouring town of Kimbolton. The *last* efforts of this kind, were prompted by his fraternal affection, which had suffered no abatement during his unusually long life: he visited his sister during her dying hours* at the close of the year 1824; and on the 11th of January, 1825, he attended her funeral,—after his return from which he never again quitted his parsonage till he himself was carried to the tomb! About a month after this he was attacked by a local disease, "not very painful," (to use his own words,) "but extremely distressing." Adverting to this complaint, in a letter to a lady connected with his family, (the last he penned), dated April 23, 1825, he observes, "I was highly gratified by your condolence on the death of my sister; Providence supported me wonderfully in health, so long as *she* lived, and my presence seemed to give her great comfort; I followed her to the grave, and since that have not been out of the house." His malady baffled the skill of his medical attendants; and he quietly breathed his last on the 3d of June, 1825, having nearly completed his 90th year.

---

* Mrs ELIZABETH LONGMIRE, relict of the Rev. Daniel Longmire, (formerly Fellow of Peter-House, Rector of Newton, Suffolk, and Vicar of Linton, Cambridgeshire,) died on the 4th of January, 1825, aged 86.—Mr. Longmire died in 1789.

The preceding pages have been devoted, almost exclusively, to an account of the *scientific and literary* pursuits of Professor Martyn; the only point of view in which a Memoir could be considered as sufficiently interesting to engage public attention. It would, however, be unjust to the memory of this excellent man, that a narrative of his life should go forth unaccompanied by some little notice of his *personal* character.

The distinguishing feature of his mind seems to have been a tranquil and even temperament, which rendered him happy in himself, and peculiarly engaging to all with whom he was connected. This quality displayed itself, according to circumstances, in general sweetness of disposition, kindness, benevolence, cheerfulness, remarkable forbearance under provocation, and abstinence from all occasions of offence to others. He was never so happy, as when he was the means of contributing to the happiness of others. A lady* of literary attainments, who enjoyed his friendship for many years, and who still lives to record the value she attached to it, observes,—

" I confess I should be sorry to see the correspondence of such a man put before the world, leaving the reader to *infer* what he was. He lives vividly in my memory. My obligations to him are many, and connected with what I may call the existence of my

* Miss Hawkins; to whom the writer of this Memoir is indebted for several obliging communications.

mind. He was the most agreeable man of science I ever knew. I do not fall in with the fashion of calling every one, with whom we are in good humour, *amiable*;—but AMIABLE, in the strictest sense, Professor Martyn was. Young as I was when he first took an interest in informing me,—and ignorant as I was,—I yet *never feared* him; though on no consideration would I have *presumed* on his kindness, which, in its outset, I did not imagine would ever have amounted to what I found it.

" He was one of those *managers of time*, who, with more to do than most people, and doing it better than any body, never want leisure, nor seem in haste. His ' Gardener's Dictionary' was a Herculean labour: but it was achieved almost *playfully*.

" When his literary life had closed, his letters * to me were invaluable. To my brother, with whom I reside, he was a classic friend; and he took up a fatherly affection for a young person, almost a stranger, whom circumstances had thrown on my care. In short, I cannot tell you what he was *not* to us. A letter, with his well known and well written superscription, excited a shout of joy, and made a feast for one or two of my neighbours."

One instance, out of many, of the forbearance of his temper, and the placidity of his disposition, may

* It was once Miss Hawkins's intention to have published a volume of this correspondence. These letters no longer exist; owing to a circumstance much to be regretted, but which it is unnecessary to record here.

here be mentioned. When residing in London, his character was grossly aspersed by the editor of one of the public prints, who rashly confounded *his* name with that of an unworthy clergyman, who had been detected in disgraceful and highly immoral conduct. A zealous friend, feeling honest indignation at this unjustifiable mistatement, vehemently expressed a wish to be allowed to give the rebuke which was merited by the careless libeller. But the Professor moderated the zeal of his warm-hearted friend; and, with a calm dignity of mind, simply remarked, that he would " send a note to the publication, and rectify the mistake."

The following passage, which occurs in a letter written in his 87th year, affords an illustration of the placid state of his mind in that advanced period of life.—Referring to a book he had received from his correspondent, he observes,—" Having had much other reading, I have not yet found time to peruse the ' *Gazette of Health.*' I am glad if you have reaped any benefit from the prescription you found in it; but I have no great confidence in Medicine, for nervous disorders. *My* ' Gazette of Health' for them, is—air and exercise; peace and tranquillity of mind; temperance and cheerful society; and the sweet comforts of genuine Christianity. Only procure these— take them morning, noon, and night,—the more the better—and, *probatum est !*"

Dated Feb. 22, 1822, and addressed to W. E. Stevens, Esq.

As he descended the vale of life, this happy tem-. perament of his mind was mellowed and matured, so as to render him one of the most agreeable companions imaginable. When considerably above fourscore, he expressed himself, to a very old friend, whose years were scarcely fewer than his own, to the following effect;—" It yields me the purest pleasure, in my old age, that I feel an *increasing* interest in the welfare of my relatives, friends, acquaintances, countrymen, fellow-churchmen, fellow-Christians, and the whole human race. I say this, to contradict the common notion, that, as age advances, old men become more morose, peevish, pettish, penurious, and selfish :—as I do not fancy myself better than others, I hope the popular opinion is not, generally, just." Whatever may be thought as to the consistency of this observation with general experience, it certainly may be considered as giving a lively picture of the venerable Professor's *own* mind; which became increasingly cheerful, benevolent, and affectionate, as he drew near to the close of his pilgrimage.

Professor Martyn's religious principles were firm and steady. While deeply conversant with the most beautiful of the works of God, in the inanimate creation, he was not forgetful of their Divine Author ; nor did he allow the pride of intellect, which is so often and so lamentably excited by considerable attainments in science, to chain down his mind to the contemplation of *second* causes. He was taught in a

s 2

better school than that of mere human reason; and had there learned to trace the finger of GOD in the order and loveliness of *his works.* He was well persuaded that the mind of a philosopher is never more truly exalted, than when abased under the conviction of the nothingness of human discovery in its utmost extent! The prevailing sentiments of his mind were those of the Christian Poet,—

> " not a flow'r
> But shows some touch, in freckle, streak, or stain,
> Of HIS unrival'd pencil. HE inspires
> Their balmy odours, and imparts their hues,
> And bathes their eyes with nectar . . . .
> Happy! who walks with HIM! whom, what he finds,
> Of flavour, or of scent, in fruit or flow'r,
> Prompts with remembrance of a present GOD!"

But while Professor Martyn's religious principles were thus confirmed and strengthened by the ennobling pursuit to which the chief part of his life had been dedicated,—his piety was by no means confined to that *devotional sentimentality* of the *mere* natural philosopher, which is sometimes mistaken for *religion.* He sincerely believed, and duly appreciated the importance of those great truths and doctrines, which Revelation alone can teach; and of which the most cultivated, and the most untutored mind are equally ignorant, until the grace of the Holy Spirit opens the understanding and disposes the heart for their reception. Nor was he ashamed to avow his principles and feelings on these points. He lamented, indeed, that in the former part of his life, he had suffered his time to be almost entirely engrossed by science; and he reflected, with regret, that the too ardent pursuit of his favourite studies, had drawn him so far within the fascinating circle of literary and philosophical society, as to leave him less leisure than was desirable, either for personal religious improvement, or for the important duties of the ministry. In short, he was humbly conscious that he had lived too much *in the world.*

This remark may call forth a sneer from those who have been themselves drawn into the giddy vortex; and who have not lived, as the subject of this memoir did, to see the vanity of the most refined and interesting pursuits, when cultivated beyond their proper limits, and when permitted to interfere with duties of more immediate and of overwhelming importance. In the latter part of his life, Professor Martyn was deeply impressed with this consideration. After the completion of his laborious Botanical work, in 1807, he ceased to devote any considerable portion of his time to scientific pursuits. Not that his *taste* for such subjects was in the slightest degree impaired; but, having reached his 72d year, he wisely reflected that an individual, who had already done so much for science, might fairly consider himself as having fully satisfied the reasonable expectations of others; and that, in the view of his

personal comfort and advantage, there were subjects
of infinitely greater interest and importance, which
invited and demanded a more undivided attention
than he had hitherto bestowed upon them. He
esteemed it a peculiar blessing, that his life was pro-
longed far beyond the period which he had devoted
too exclusively to science; and that he was favoured
with strength, for many years, to preach to his
beloved flock at Pertenhall, those great truths which
were the stay and the solace of his declining age.
He was in the habit of occupying his own pulpit (with
few exceptions, when prevented by ill health), until
his 82d year; and on these occasions the truly venerable preacher delivered his message with much
earnestness and affection.

"My removal to Pertenhall in 1798," he observes,
in some private notes of his own life, written in his
86th year, "was a providential one for me; and I
humbly hope has proved a blessing to me, under
Divine Grace, by weaning me from the world; in the
business and pleasure of which I had been too much
involved, during the last 25 years. The peaceful and
quiet retreat of this village has contributed to the
health of my body, and has brought my mind into a
frame gradually more fitted for my final removal;
which has been put off, by Divine favour, to a far more
distant period than I could ever have expected, and
has been graciously unaccompanied with that labour
and sorrow, which commonly attends human life after

the term of four-score years. I have often said, with
the heathen poet,—

Ελπις και συ Τυχη μεγα χαιρετε, τον λιμεν' ευρον!

Inveni portum ! Spes et fortuna valete !·
Sat me lusistis, ludite nunc alios.

Fortune and hope, farewell! I've found the port!
With me no longer, but with others sport.

—" or rather, to express myself in language more appropriate to the Christian,—

To fortune and the busy world, farewell!
With Faith and heav'nly Hope, O let me dwell!

" Sensible of a gradual decay, both in my bodily
powers, and in the faculties of my mind; but grateful
to God for such remains of them as enable me to
read, to correspond with my distant friends, and to
enjoy the society of my dear relations; I am waiting
in awful expectation, though in firm *faith*, and in
cheerful *hope*, for my final call to appear in the presence of my God and Saviour! My tranquil state,
however, has not been without its afflictions. Besides
the loss of many friends,—a loss incident to all who
live to old age,—I have had to weep over the un-
timely deaths of four lovely and pious female relatives,
in the prime of life. . . . . It is a mournful catalogue! But it is good for *them* that they are gone to
the bosom of their Saviour and their God! It is

good for my family, and for *me*, that we have been afflicted! Blessed be God for all his mercies! "

It was during one of those seasons of affliction, alluded to in the preceding paragraph, that the writer formed an acquaintance with the venerable subject of this Memoir; having been induced to take the temporary charge of the Professor's parish, in the summer of 1815, when death had visited his son's family under circumstances peculiarly distressing. This engagement (though short) gave him an opportunity of personally observing the delightful character he has just been drawing; and it providentially laid the foundation of a continued intercourse with the Professor's family, which some years after happily determined the the most important step in his life.

Professor Martyn was a warm supporter of the British and Foreign Bible Society; and on no occasions did he manifest more pious and affectionate feelings, or a more Catholic and charitable spirit, than when his flock, and his religious friends of various denominations, were annually gathered around him to promote the success of that noble Institution. When health permitted, he always presided at these interesting meetings; and the patriarchal simplicity of manners, and the unaffected devotion of the aged Rector, diffused a spirit of love among the people, which is remembered with thankfulness and delight by many who enjoyed the privilege of attending these anniversaries. It was his habit to open these meetings by prayer; for which he prepared notes, several of which are before the writer,—the last of them being endorsed " *Oct.* 20, 1824, *Anno Ætatis nonagesimo !*" These afford a beautiful picture of the religious state of his mind, at the very close of his days; for he lived only eight months after the period just mentioned.

The circumstances of his death have already been noticed (p. 255). He was interred in the Chancel at Pertenhall, where the following Epitaph is inscribed upon a marble tablet, surmounted by the family arms;—(viz. *Argent, two Bars Gules ;* Crest, *a Leopard's Head coupé proper ;*)—

In Memory of
THOMAS MARTYN, B.D.,
Professor of Botany
In the University of Cambridge,
and
Rector of this Parish.

Having distinguished himself in the literary world by many useful publications, he spent his latter years in this retired village in preparing to meet his God : and, in the exercise of benevolence to the poor, condescension, patience, and kindness to all, he adorned the Christian character, ripened for heaven, and lives in the hearts of those who knew him.

He was born Oct. 4, 1735;
and died June 3, 1825,
in the 90th year of his age,
Like as a shock of corn cometh in, in his season ! JOB. v. 26.

There are the following engraved portraits of Professor Martyn:—

1. By J. Farn, from a painting by S. Drummond, December, 1796, Æt. 61, 8vo.

2. By Vendramini, from an excellent painting by Russell, R. A., (now in the possession of the Rev. J. K. Martyn,) June, 1799, Æt. 63, folio. This portrait is a striking likeness; the engraving is ornamented with a figure of the *Martynia proboscidea,* or *Horn-capsuled Martynia.*

3. By Holl, from the above-mentioned painting by Russell, 1799, Æt. 63, 8vo. This engraving is better executed than that by Vendramini, and the likeness is equally well preserved.

LIST OF THE PRINTED AND MANUSCRIPT WORKS OF PROFESSOR THOMAS MARTYN.

I. *Published Works.*[a]

1. Plantæ Cantabrigienses; or a Catalogue of the Plants which grow wild in the County of Cambridge.—Herbationes Cantabrigienses, or Directions to the places where they may be found.—Lists of the more rare Plants growing in many parts of England and Wales. 8vo. 1763.

2. Heads of a Course of Lectures on Botany. 8vo. 1764.

3. A List of the most remarkable Weeds in England. Mus. Rustic. Vol. V., No. LVI., for 1765. (P. B. C.)

4. Descriptions of the most remarkable annual Weeds in England, with 15 figures. Mus. Rustic. Vol. VI., No. XXVIII., for 1766. (P. B. C.)

5. Descriptions of the most remarkable biennial and perennial Weeds. Mus. Rustic. Vol. VI, No. LXIV., for 1766. (P. B. C.)

6. The English Connoisseur; containing an account of Paintings, Sculpture, &c., in England. 2 vols. 12mo. 1766. (Anonymous.)

7. A Sermon preached at Great St. Mary's, Cambridge, for Addenbroke's Hospital. 4to. 1768.

8. Some Account of the late John Martyn, F. R. S., and his Writings. 12mo. 1770.

---

[a] [The titles are here given in a compendious form; as a more full account of each work may be seen in the preceding Memoirs under the respective date of publication.]

9. A Chronological Series of Engravers. 12mo. 1770. (Anonymous.)

10. Catalogus Horti Botanici Cantabrigiensis. 8vo. 1771.

11. Mantissa Plantarum Horti Botanici Cantabrigiensis. 8vo. 1772.

12. The Antiquities of Herculaneum, translated from the Italian. Vol. I., Part I., containing the Pictures. (Published in conjunction with Dr. Lettice.) 4to. 1773.

13. Elements of Natural History. Part I. Mammalia. (Not continued). 8vo. 1775.

14. Heads of a Course of Lectures in Natural History, read at the Botanic Garden, Cambridge. 12mo. 1782.

15. An inquiry into the nature and use of Pozzolana Earth. Printed for the Transactions of the Cambridge Society for the promotion of Philosophy, &c., but not published. 4to. 1785.

16. Suggestions on the utility of publishing a Catalogue of Plants with the names accented. Gent. Mag., Vol. LV., Part II., p. 757. 8vo. 1785. (P. B. C.)

17. Letters on the Elements of Botany, translated from Rousseau; with 24 additional Letters. 8vo. 1785.

18. Thirty-eight Plates, with explanations, to illustrate Linnæus's System. 8vo. 1788.

19. A Tour through Italy. (Anonymous.) 12mo. 1787.— An Appendix to the Tour through Italy. (Anonymous). 12mo. 1787.

20 Sketch of a Tour through Switzerland. (Anonymous). 12mo. 1787.—An Appendix to the Tour through Switzerland; with an account of Saussure's Ascent of Mont Blanc. (Anonymous). 12mo. 1788.

21. A Tour through Italy, &c., &c. Considerably enlarged. 2d edition, (with the Author's name). 8vo. 1791.

22. Observations on the Language of Botany. Trans. Lin. Soc., Vol. I., pp. 147—154., 4to. 1791.

23. The Language of Botany; being a Dictionary of the terms made use of in that Science. 12mo. 1793.

24. Flora Rustica; exhibiting figures of such plants as are either useful or injurious in Husbandry. 4 vols. 8vo. with 144 coloured plates. 1792—1794.

25. An account of the Natural History of Little Dalby, Leicestershire. Nichols's Leicestershire, Vol. II., Part I., pp. 160—162. Folio. 1798.

26. A description of the *Hæmanthus multiflorus*, or Blood-flower. With a coloured plate. 8vo. (1795.)

27. Observations on the flowering of the *Anagallis arvensis, Œnothera biennis,* and *Hibiscus trionum.* Trans. Lin. Soc. Vol. IV., pp. 158—163. 4to. 1797.

28. The Gardener's and Botanist's Dictionary; by the late Philip Miller, F. R. S. With a complete enumeration and description of all plants hitherto known, their culture, &c., &c. 4 vols. (titled as 2 vols. in two parts.) Folio. 1807.

29. Observations on Grasses, supplementary to those of Stillingfleet.—Observations on the times of leafing, flowering, and the fall of the leaf, of certain plants and trees, from the year 1775 to 1809: arranged in tables.—Printed in the Miscell. Tracts of Stillingfleet, edited by the Ven. Archdeacon Coxe; Vol. II., Part II., pp. 305—357. 8vo. 1811.

30. A List of rare plants in Surrey. Printed in Manning's History of Surrey, by Bray; Vol. III. pp. lxv—lxx. Folio. 1814.

Professor Martyn occasionally contributed to the periodical publications, either anonymously, or under the signiture M. T., and P. B. C.

His communications to the *Museum Rusticum*, 1795—6, and to the *Gentlemen's Magazine*, 1788, have been noticed already.

The *British Critic* for 1796, has several articles from his pen.

Analytical Review, 1790, Vol. III., p. 68., (M. T.)—Review of Withering's Arrangement of British Plants, Vol.III. Part I.

Analytical Review, 1791, Vol. IX., p. 192., (M. T.)—Review of Wildenow's Historia Amaranthorum.

Analytical Review, 1791, (M. T.)—Review of Pulteney's Sketches of the Progress of Botany in England.

Analytical Review, 1792, Vol. XIV., p. 423. (M. T.)—Review of Withering's Arrangement of British Plants, Vol. III., Part. II.

To *Nichols's Literary Anecdotes of the* 18*th Century*, he furnished, in 1814,—

1. Some Account of Mr. Montagu Bacon. (Vol. VIII p. 417.)

2. Some Account of Dr. Farmer. (Vol. VIII., pp. 420—421.) See above, p. 97.

3. A correction of Mr. Nichols's mis-statement that he was the Author of " Aranei, or the Natural History of Spiders," a work written by Thomas Martyn, the Entomologist, of Marlborough Street, a native of Coventry. (Vol. VIII. p. 432.)

II. *Manuscripts.*

1. Travels through Switzerland and Italy, during the year 1778, 1779, 1780. Two vols. 4to. pp. 1040.

2. An Auto-biographical Account of himself, drawn up in his eighty-sixth year. 1821. 4to. pp. 53.

———

The Autograph of Thomas Martyn. Æt. 55, 1790.

INDEX.

T

U

GAULTER, Printer, Lovell's-Court, Paternoster-Row.

ERRATA.

Page 9. line 31, *for* 1824,) *read* 1824,
—— 9, —— 32, *for* Museum *read* Museum)
—— 15, —— 18, *for* Massay *read* Massey.
—— 76, —— 22, *for* Anee. *read* Anec.
—— 76, —— 29, *for* Botanicum *read* Botanicum.
—— 87, —— head-line, *for* Æt. 13 *read* Æt. 14.
—, 91, —— 14, *for* kind of *read* kind.
—— 98, —— 17, *for* Methodists *read* Methodists ᵃ.
—— 116, —— 26, *dele* and read.
—— 119, —— 21, *for* Botanical excursions 13. *read* 13 Botanical
excursions.
—— 137, —— 19, *for* circumjectili *read* circumjectili.
—— 193, —— 1, *for* againe *read* again.
—— 193, —— 2, *for* hav *read* have.
—— 214, —— 23, *for* Battata *read* Batata.
—— 214, —— 24, *for* Battatas *read* Batatas.
—— 221, —— 9, *for* pp. 50, —, *read* pp. 50, 180
—— 232, note (ᵃ) *should be included in hooks.*
—— 267, line 26, *for* date *read* dates.
—— 270, —— 4, *for* 1795 *read* 1765.

CPSIA information can be obtained at www.ICGtesting.com
Printed in the USA
LVOW11s1923010414

379826LV00019B/865/P